GENE TECHNOLOGY AND THE PUBLIC

Gene Technology and the Public
An Interdisciplinary Perspective

Edited by Susanne Lundin and Malin Ideland

NORDIC ACADEMIC PRESS

NORDIC ACADEMIC PRESS
BOX 935
S-220 09 LUND
SWEDEN

© 1997 NORDIC ACADEMIC PRESS AND THE AUTHORS
ISBN 91-89116-00-3
PRINTED IN SWEDEN
TEAM OFFSET MALMÖ 1997

Contents

Foreword

Genetic knowledge in combination with recombinant-DNA techniques have proved to be powerful technological instruments in many sectors of society. Gene technology is a valuable method for plant and animal breeding in agriculture, forestry and fishery. It is indispensable for the pharmaceutical industries and it has provided new diagnostic tools in health care. Gene therapy protocols have been officially approved for some diseases, and with the rapid development of the mapping and sequencing of the human genome we will soon see proposals for gene therapy for many monogenetic and polygenetic diseases. Genetic testing is an established part of forensic medicine, and regulations are requested for the implementation of such tests in workplaces and by insurance companies.

In many cases the implementation of gene technology challenges established social values and attitudes. Discussions in parliaments and governmental committees and debates in the mass media indicate that many people are morally concerned about what is happening in this field. The creation and introduction of transgenic animals and transgenic food products is sometimes strongly resisted by non-governmental organizations. The implementation of gene technology has in many cases proved to be a touchy issue. It is important to be sensitive to public perceptions in order to create appropriate guidelines that are publicly acceptable. It is also important that estimates about public perceptions with regard to gene technology are based on established scientific knowledge and supported by research.

In order to identify and document a broad spectrum of individual responses to the implementation of gene technology an interdisciplinary approach is required. Values, attitudes, beliefs and world views are concerned, as well as the social impact on a larger scale.

Scholars from different scientific fields must cooperate in order to arrive at a comprehensive understanding of the impact of gene technology on public perceptions. Neither gene technology in itself nor its application in different areas knows any geographical boundaries. The question about the impact of gene technology on public perception has a strong European dimension, and common regulations are required for the members of the European Union. Research concerning the public perception of gene technology is much needed but is as yet only in a preliminary stage of development.

The expert workshop 'Public Perceptions of Gene Technology' invited scholars in several disciplines from different European countries, the purpose being:

1   to give presentations of research already in progress concerning public perception of gene technology;
2   to present results from research in other areas that can be of interest for future projects concerning the public perception of gene technology;
3   to identify problems and topics that can be approached in multidisciplinary approaches in this area;
4   to initiate a network of scholars from different disciplines and European countries whose concerted efforts can help to improve our understanding of the public perception of gene technology and its implementation in different sectors of society.

The workshop was held at Lund University, Sweden, 6–7 February 1997 and was jointly hosted by the Department of Ethnology and the Department of Medical Ethics at Lund University together with the ELSA-group of the Swedish national research councils.[1] The workshop was financially supported by The European Commission, Lund University and the ELSA-group.

Mats G. Hansson
Coordinator ELSA-Sweden

Note

1 The ELSA-initiative was taken in 1995 by several research councils and other Swedish grant agencies: The Swedish Cancer Society, The Swedish Council for Planning and Coordination of Research (FRN), The Swedish Natural Science Research Council (NFR), The Bank of Sweden Tercentenary Foundation, The Swedish Research Council for Engineering Sciences (TFR), The Swedish Council for Forestry and Agricultural Research (SJFR), The Medical Research Council (MFR), The Swedish Council for Social Research (SFR), The Swedish Council for Work Life Research (RALF), and HSFR (The Swedish Council for Research in the Humanities and Social Sciences).

Genetics, Genetic Engineering, and Everyday Ethics
An Introduction

Susanne Lundin and Lynn Åkesson

Today biology occupies a central place in people's consciousness. If the 1970s were the age of social explanatory models, when issues and problems in society were regarded from the perspective of social psychology, the 1990s have given primacy of interpretation to biology and genetics. Scholarly debates today are investigating the significance of the genetic inheritance for our social existence (Haraway 1992; Butler 1993). The fact that many people are inclined to let genetics be the interpretative framework imposed on reality is evident not only in the public debate led by the experts. On a more everyday level, people's lives are increasingly pervaded by a flow of biomedical knowledge showing the associations between biology and identity. This is obvious in many arenas. For example, staff in maternity care emphasize the biological effects of breast-feeding above the emotional importance; pupils' learning problems are no longer interpreted in social but in genetic terms; children reading comics or playing video games are introduced to the world of genetically mutated beings.

The biomedical knowledge that pervades society has consequences on several levels. From a scientific perspective this means that new insights can be applied in medical practice. At the same time, there are also changes in people's cultural practice. In the wake of genetic

engineering, new practices and ideologies arise which have consequences not just for individuals but also at a collective societal level
(Edwards 1993; Marteau and Richards 1996; Price 1995). Today
people feel that, although the body is a complicated apparatus, it
can still be transilluminated, measured, and treated. More than ever
before, the body appears to be a Pandora's box, an ingenious construction containing both possibilities and threats. This often con-
flict-filled awareness has consequences for people's practice, for the
norms, values, and patterns of action that emerge in society.

There is a tendency to interpret human questing and existential
insecurity as expressions of modernity (Giddens 1991). Although
there is a great deal to suggest that this is correct, it should nevertheless be stressed that similar expressions can be found in widely
varying eras. People's fears of the consequences of genetic engineering, or rather the way in which these fears are expressed, are very
similar to reactions to the unknown in nineteenth-century peasant
society.[1] The desire to change and influence the possibilities of the
body is not new. What is new is the technology by which the desire
can be fulfilled. A historical comparison may reveal what is taken
for granted in our own times, thus disclosing the cultural context
of the natural sciences (cf. Kemp 1991; Lundin and Åkesson 1995;
Johannisson 1994).

The Encounter of Different Conceptual Worlds

In the natural and human sciences, important questions are being formulated today about how we should deal with genetic engineering. The very speed of medical development makes an interdisciplinary discussion important (Stansworth 1987). This is particularly clear in the growing research dialogue between scientists
and humanists. At the same time, there is an increased will to bring
the debate to the general public, to give ordinary people a chance
to see what is happening in technology and also to assume responsibility for it (Nelkin and Lindee 1995; Rabinow 1996; Uddenberg 1996). This attitude is clearly seen in the increasing number
of articles in the press and information programmes on radio and
television – not least those aimed at the young. The scientific di-

alogue is characterized by a desire to bridge the differences that exist between the world of the experts and that of the general public. Yet these efforts often run into problems. The discussions bring up concepts such as human dignity and co-determination, almost as a contrast to concepts such as responsibility and expertise (Rose 1994). It is claimed that information about genetic engineering is inaccessible, but that this information is needed if people are to assume responsibility for the consequences of technology (SOU 1989:75; SOU 1992:82).

This shows the importance of not just spreading information and conducting opinion polls to establish people's attitudes, in other words, implementing almost ethnocentric enlightenment projects. If one wants to break down barriers to understanding, what is needed is not just serious specialist discussion but also a willingness to take all the voices seriously (Lundin 1995). It goes without saying that the layman is also involved in scrutiny and ethical debate; this popular critical inspection of the new technologies is partly due to the bewilderment about how they should be handled in everyday life.

This popular debate, however, is not conducted in a way that the scientists would regard as a serious discussion. Instead it is visible in a specific narrative structure: in newspaper headlines, in letters to the editor, and in the way people talk about their lives and the society around them. It is in such contexts that 'truths' are formulated, such as the claims that nice-looking but tasteless carrots must have been genetically manipulated, or that twin births must be a sign of test-tube fertilization, or that *in vitro* fertilization offers the chance of having a 'custom-made' child. There are obviously different knowledge systems which live parallel to each other or integrated with each other, but which are defined and interpreted in very different ways. This reveals the complexity and inherent problems that characterize today's society.

Parallel to the interest of the experts and the public in investigating the potential of genetics, there is a seemingly opposite desire – an ambition to tone down the potential of genetic engineering, its power for medical and societal change. Instead, the deterministic perspective of genetics is stressed, that is, the idea that everything is already biologically determined and therefore beyond social control (Lundin 1996). In the media there are more and more articles shifting the responsibility from society to biology; we read headlines such as 'Bedwetting is hereditary', 'Your thumb can show whether you are homosexual', 'Alcoholism is in the genes' (cf. Dawkins 1976; Ideland [Svantesson] 1995; Jones 1996; Ideland 1996).

Apart from the truth or otherwise of the headlines, they appear to contain a longing to escape responsibility. It is undoubtedly difficult, both for the individual and for society, to shoulder the burden of deciding how we should use genetic information (cf. Hermerén 1993; Anderberg, Jeffner, and Uddenberg 1994). Freedom of choice requires taking responsibility, and the longed-for freedom quickly becomes the opposite.

Public Perceptions of Gene Technology

The problematic of genetics and genetic engineering refuses to be limited by subject boundaries. Putting the whole person at the centre is a necessary perspective for natural, social, and cultural sciences alike (Åkesson 1997). When searching for knowledge about the associations between health and identity especially, it is important to analyse both biology and culture.

Until now there have been, both nationally and internationally, only few scientists and culture analysts with any interest in interdisciplinary comparisons. In recent years, however, the natural and cultural sciences have begun to come closer together.

At the Department of European Ethnology at Lund University, such a collaboration is in progress through the project 'Genethnology: Genetics, Genetic Engineering, and Everyday Ethics'.[2] The overall aim is to study the social and cultural consequences of the rapidly expanding fields of biomedicine from an interdisciplinary perspective including scholars from ethnology and genetics. Another aim of the project is to develop a creative and expansive research environment. One attempt for such a common dialogue where unexpected and innovative knowledge might arise was the European expert workshop 'Public Perceptions of Gene Technology', held in Lund on the 6th and 7th of February 1997. It was organized by the Department of Ethnology in association with the ELSA-group in Sweden and the Department of Medical Ethics at Lund University. The purpose of this book is to document the presentations given at the workshop and to show different epistemological traditions concerning public perceptions of genetic engineering.

The outline of *Gene Technology and the Public: An Interdisciplinary Perspective* mirrors the topics discussed at the workshop. Part one, 'Public Perceptions from a Multidisciplinary Perspective', is composed of three articles where individual and public experiences of biomedicine and genetic engineering are in focus. From the theological perspective Carl Reinhold Bråkenhielm and Katarina Westerlund discuss the role of biology and genetics in particular as a producer of ideology. Similar themes are brought up from an anthropological viewpoint and with the focus on individuality by Elini Papagaroufali, who examines people's attitudes to organ donation and xenotransplantations. A quite different presentation, grounded on psychological and statistical methods, is given in Lynn Frewer's discussion of consumer acceptance of genetically modified food.

Part two, 'Everyday and Clinical Experience of Gene Technology', deals with cultural and clinical experiences of genetic engineering.[3] The culture sociologist and psychoanalyst Alberto Melucci takes up the fact that people today have developed the reflexive capacity, as he says, to produce their own reproduction and the environment itself. This increased intervention in our inner Nature and the Nature surrounding us, takes the form of a pure symbolic capacity allowed by science, as today's gene technology testifies. Also the ethnologist Susanne Lundin discusses cultural aspects and puts the question of how new medical technology influences people's lives. Using biographical as well as public media material, she illustrates how biomedicine and gene technology is integrated in people's everyday lives, and how values and norm systems are affected. Another ethnological perspective is given by Malin Ideland. In her presentation she analyses film as a folkloristic genre commenting on genetic engineering. From a clinical point of view, Maria Anvret, from the Department of Neurogenetics, points out that new technology can easily detect mutations, such as cancer or Huntington's chorea, which means that risks of inheriting genetic diseases can be calculated. But how should this knowledge be used? In other words: how does gene technology influence the potential of medical genetics?

The subject of part three is 'Social Sustainability of Gene Technology'. The four contributors have similar starting-points – genes and social values. The sociologist Torben Hviid Nielsen presents results from the Eurobarometer and discusses whether knowledge of biotechnology improves acceptance figures. Björn Fjæstad, Science Communication, likewise deals with values of gene technology and the formation of public opinion. The purpose of his study is to identify ethical positions of decision-makers with regard to concerns about animals, plant life, and ecosystems. The European Values Study, as the sociologist Loek Halman shows, is designed to explore basic values in the domains, as for example religion, morality, politics, family, and work. In his article he addresses not only the concept of values, but also the measuring of values and the establishment of cross-national comparability. Wim Heijs and Cees Midden's contribution, finally, discusses from a psychological perspective attitudes towards biotechnology based on a survey programme comprising a series of four national studies in the Netherlands.

With *Gene Technology and the Public: An Interdisciplinary Perspective* we hope to highlight some of the problems that must be dealt with in the discussion of what the new genetics means to people – experts as well as laypersons. We hope this book will contribute to a continued interdisciplinary dialogue among scholars and others with an interest in modern genetics.

Notes

1   Work on our now finished project 'Transformations of the Body' has shown unexpectedly tenacious structures in people's attitudes to the manipulation of what is biologically 'natural' or 'God-given' (Lundin and Åkesson 1996a; see also Merchant 1989; Stattin 1990).

2   The initiative for the research group at Lund University has been taken by Susanne Lundin and Lynn Åkesson of the Department of European Ethnology. The group includes Malin Ideland, doctoral student in ethnology, Ingrid Frykman, Depart-

ment of Genetics, and Ulf Kristoffersson, Department of Clinical Genetics. See
Lundin and Åkesson 1996b; Hansson 1996.
3   Gunnel Elander from the department of Medical Ethics, Lund University, also
took part in this session. Her contribution 'Diagnostics without cure or treatment'
is published elsewhere.

References

Åkesson, Lynn. 1997. *Mellan levande och döda: Föreställningar om kropp och ritual.* Stockholm: Natur och Kultur.

Anderberg, Thomas, Anders Jeffner, and Nils Uddenberg. 1994. *Biologi och livsåskådning.* Stockholm: Natur och Kultur.

Butler, Judith. 1993. *Bodies That Matter: On the Discursive Limits of 'Sex'.* New York and London: Routledge.

Dawkins, Richard. 1976. *The Selfish Gene.* Oxford: Oxford University Press.

Edwards, Jeanette. 1993. Explicit Connections: Ethnographic Enquiry in North-West England. In *Technologies of Procreation: Kinship in the Age of Assisted Conception*, ed. Jeanette Edwards, Sarah Franklin, Eric Hirsch, Frances Price, and Marilyn Strathern. Manchester: Manchester University Press.

Giddens, Anthony. 1991. *Modernity and Self-Identity: Self and Society in the Late Modern Age.* Cambridge: Polity Press.

Hansson, Mats G. (ed.). 1996. *Report of Research concerning Ethical, Legal and Social Aspects of Genome Research: ELSA-activities in Sweden.* Stockholm: FRN.

Haraway, Donna. 1992. When Man is on the Menu. In *Incorporations*, ed. Jonathan Crary and Sanford Kwinter. Zone 6. London: MIT Press.

Hermerén, Göran. 1993. *Tankar om Gen-Etik.* Stockholm: Hybrid DNA-delegationen.

Ideland [Svantesson], Malin. 1995. Genteknik och vardagsetik: Arbetsrapport. Department of European Ethnology, Lund University. (mimeo)

Ideland, Malin. 1996. Hormoner, gener och kulturella konstruktioner. Seminar paper, research school in historical anthropology, Lund University.

Johannisson, Karin. 1994. *Den mörka kontinenten: Kvinnan, medicinen och fin-de-siècle.* Stockholm: Norstedts.

Jones, Steve. 1996. *In the Blood: Good, Genes and Destiny.* London: Harper Collins Publishers.

Kemp, Peter. 1991. *Det oersättliga: En teknologietik.* Stockholm: Symposion.

Lundin, Susanne. 1995. Är provrörsbefruktning naturligt? *Ordfront Magasin* 1995:3.

Lundin, Susanne. 1996. När blicken för livet vänds inåt. In *Samtider, Svenska Dagbladet,* 1 March 1996.

Lundin, Susanne, and Lynn Åkesson. 1995. Att skapa liv och utforska död. *Kulturella Perspektiv* 1995/1.

Lundin, Susanne, and Lynn Åkesson (eds). 1996a. *Bodytime: On the Interaction of Body, Identity, and Society.* Lund: Lund University Press.

Lundin, Susanne and Åkesson, Lynn. 1996b. Creating Life and Exploring Death. *Ethnologia Europaea* 26:1.

Marteau, Theresa, and Martin Richards. 1996. *The Troubled Helix: Social and Psychological Implications of the New Human Genetics.* Cambridge: Cambridge University Press.

Merchant, Carolyn. 1989. *The Death of Nature: Woman, Ecology and the Scientific Revolution.* San Francisco: Harper.

Nelkin, Dorothy, and Susan Lindee. 1995. *The DNA Mystique: The Gene as a Cultural Icon.* New York: W. H. Freeman and Company.

Price, Frances. 1993. Beyond Expectation: Clinical Practices and Clinical Concerns. In *Technologies of Procreation: Kinship in the Age of Assisted Conception,* ed. Jeanette Edwards, Sarah Franklin, Eric Hirsch, Frances Price, and Marilyn Strathern. Manchester: Manchester University Press.

Rabinow, Paul. 1996. *Making PCR: A Story of Biotechnology.* Chicago: The University of Chicago Press.

Rose, Hilary. 1994. *Love, Power and Knowledge: Towards a Feminst Transformation of the Sciences.* Cambridge: Polity Press.

SOU 1989:75. *Etisk granskning av medicinsk forskning: De forskningsetiska kommittéernas verksamhet.* En underlagsstudie

från Forskningsetiska utredningen. Statens offentliga utredningar 1989:75. Stockholm.

SOU 1992:82. *Genteknik: En utmaning.* Betänkande av
Genteknikberedningen. Statens Offentliga Utredningar
1992:82. Stockholm.

Stansworth, Michelle. 1989. *Reproductive Technologies: Gender,
Motherhood and Medicine.* Cambridge: Polity Press.

Stattin, Jochum. 1990. *Från gastkramning till gatuvåld.* Stockholm: Carlssons.

Uddenberg, Nils. 1996. *Det stora sammanhanget: Moderna
svenskars syn på människans plats i naturen.* Nora: Nya Doxa.

Worldviews and Genetics

Carl Reinhold Bråkenhielm and Katarina Westerlund

The Swedish economist Gunnar Adler Carlsson published his 'Manual for the Nineties' in 1990. In this book he argues for a form of genetic determinism. In many – if not most – areas where human beings think they are free, they are really determined by powers beyond their control. We think that our behaviour is the result of intellectual deliberations and rational calculations, whereas in reality our actions should be explained in terms of genetics and biology. Genetic factors are – says Adler Carlsson – more important than cultural or environmental factors. He calls this a belief in a *weak genetic determinism* (Adler Carlsson 1990: 118). In less enlightened times human beings believed in fairies and gnomes, devils and witchcraft. Today we know better. Genes and segments of DNA are the true causes of human behaviour. 'They are as invisible as witchcraft for the naked eye. But their power can be as great, far greater. And real!' (Adler Carlsson 1990: 74.)

This is not the place to discuss the various forms and merits of genetic determinism. Rather, we want to present a research project concerning attitudes to genetics and genetic technology among ordinary men and women in Sweden today. Twenty-four persons have participated in one-hour interviews concerning their thoughts in this area. In the second half of this paper we shall present some preliminary results of our analysis of these interviews. In the first half the general approach and the theoretical framework of the study will be described.

Worldview as Conceptual Horizon

One of the main ideas of the study is to explore the role of world-views in the formation of public attitudes to genetics and genetic technology. Gunnar Adler Carlsson's belief in weak genetic determinism could be described as a worldview. His belief could be described as an aspect of his conceptual horizon in terms of which he wants to bestow meaning and significance to human life. Such a conceptual horizon is often closely linked to a system of values and norms. The central part of this system can be described as an answer to the question: what is most important and valuable in the world? Anders Jeffner – a theological scholar at Uppsala University – says that person's worldview and central value system is linked to a basic mood of e.g. optimism or pessimism, trust or despair. Jeffner gives the following example:

> One can easily see that these three factors are significant both in all religions and in many philosophical systems. A Christian has a basic value-pattern characterized by the Christian commandment to love God and neighbour. A Christian sees the whole of reality as a creation by God in which mankind has a specific and important position. The basic mood for a Christian – at least according to many theologians – is hope. But the question that concerns us at present is: What happens to these valuations and theories when Christianity loses its importance (Jeffner 1992: 138)?

To find an answer to this question, various studies have been conducted to explore views of the world among Swedish people in general. We shall not go further into the results of these studies (some results are presented in Jeffner's article), but one thing is important in the present context. Many people suggest a biologistic view of the world. The conceptual horizon is the ecological system. Sometimes it leaves room for human freedom, but some people come close to a more deterministic idea not far from Gunnar Adler Carlsson's belief in a weak form of genetic determinism.

How can this biologistic view of the world be analysed? One possibility is the following. Human beings are meaning-seeking creatures. We try to make sense of particular impressions and – more generally

– of the various domains in which (post)modern life is differentiated (Halman and Pettersson 1995). Some people also seek a particular pattern in reality in terms of which their experiences and their knowledge can be understood. The metaphysical systems of Western philosophy can be described as such efforts to make sense of human experience and deliver us from the predicament to experience the world as 'just one thing after another'. Some people seem to dislike this intensely, while others don't seem to be bothered by it.

How do such general patterns of reality arise? Well, often a more limited fact or event is exalted to a more general model and applied to the overall reality. Plato saw the empirical world as a shadow of a higher reality, and the Biblical tradition has its basic model taken from the personal sphere and human creation. Since the beginning of modern science, however, the most important models for worldview construction have come from physics. Today it seems that this situation is changing. Biology has emerged as a supplier of models and metaphors which are generalized to serve as worldviews. Ecology is one example of this, but genetics is also important to make more general sense out of human actions and human behaviour. Genetic determinism can be analysed in this way.

In sum, modern biology delivers material for the construction of worldviews, which in turn affects our perception of human nature. One aspect of this process has been studied by Evelyn Fox Keller in her book *Refiguring Life* (1995). She looks at the discourse of the life sciences and how the scientific concept of a gene evolves into the metaphor of the gene where it is endowed with agency, autonomy and causal primacy. This metaphor in turn direct the search for evidence and defines research agendas.

The way biological knowledge is transformed into a worldview is also expressed on the cover of a recent book on the cultural impact of genetics (Figure 1). The picture suggests that the gene or the DNA-molecule is bestowed with a meaning and a significance, which exalts it above everything else in human nature. A certain answer is suggested to the question of which meaning system is to be preferred in the description and explanation of human behaviour. The answer is: the meaning system of biology and genetics. Sociological, psychological and personalistic meaning systems are

Figure 1. The cover of a recent American book on the cultural impact of genetics.

pushed into the background – not to mention theological ones. Nevertheless, certain magical or religious feelings seem to be relocated to a more immanent object. We shall return to this point.

Worldviews and Public Perceptions of Genetics

What is the significance of these reflections for the study of public perception of genetics and genetic technology? Are similar processes at work in the formation of 'worldviews'? This is, of course, a matter of empirical evidence, but a reasonable working hypothesis might be found if we contrast the formation of worldviews out of biological facts and concepts with *the opposite influence of previously acquired worldviews upon the interpretation of genetics/genetic technology.* Religious beliefs, political ideologies, ideas of nature and central ethical values and norms might lead people in general to perceive modern biology and its applications in a particular way. We have assumed that this kind of relationship is more common among people in general than the opposite, which might be a fruitful working hypothesis when studying arguments among philosophers and scientists.

One particular example of how a previously acquired worldview affects the interpretation of genetic technology is found on another cover of a Swedish anthology on genetics (Figure 2). The myth of the fall is brought into play to bring about an attitude of caution, if not abstention. The example underlines three things of more general importance.

First, it is important to acknowledge the fact that worldviews, religious or secular, are often expressed in the form of myths, metaphors and symbols. Worldviews come to us in *symbolic forms.* Religion, ideology and secular worldviews are wrapped in myths, images and stories. These symbolic forms are open to new interpretations. Figure 2 is an example of this. Nature might be interpreted as Mother Earth or in the symbol Gaia, human beings as naked apes, or survival machines for our genes. These pictures play an important role in so far as or when they are linked to modern biology.

Secondly, there are also certain central concepts of religious studies, which might illuminate the public perception of genetics. Take the concept of taboo. It is a Polynesian expression, which has come

Figure 2. The cover of a Swedish book on ethics and genetic engineering.

to be used by anthropologists to describe an intricate social phenomenon. Originally, taboo has both a positive and a negative aspect, which makes it parallel to Rudolf Otto's concept of the holy. It attracts but at the same time it arouses fear. Taboo has come to be associated with the negative, forbidden and fearsome. All cultures – including modern culture – have certain taboo in the form of both persons and objects. The gene might sometimes come close to a taboo. Genetically modified food is shunned in a way that goes beyond what might be expected on purely rational grounds.

Thirdly, there are important questions concerning the relationship between the worldviews and values/norms. Worldviews might sometimes undergird our values, but certain values are, as it were, very loosely connected to a particular worldview. Take, for example, the idea of the supreme value of human beings, or the idea of human rights. Historically they might have a certain link to Christian belief, but today it is clear that they have largely lost this relationship. Other values/norms are more closely related to a certain worldview, such as the norm that human beings must not play God or the idea that survival value is the ultimate moral value.

What is the importance of these concepts and general observations for the analysis of public interpretations of genetic technology? How far are attitudes to genetics shaped by previously acquired worldviews? What is the importance of more general attitudes to nature and technology? In what way do convinced religious believers differ in their attitudes – and why? A study designed to contribute to an answer to these questions is presently being conducted in the context of a larger research project led by Assistant Professor Nils Uddenberg at the Institute of Future Studies, Stockholm. Katarina Westerlund has completed 24 one-hour interviews with nurses, social workers, farmers and biology teachers. We have not come very far in the analysis of these interviews, but Katarina Westerlund has some preliminary remarks on the highly interesting material.

The Interview Study

The understanding of worldviews in the study has followed Anders Jeffner and, as briefly mentioned above, we understand a worldview as consisting of theoretical convictions of the human being and the world as a whole, including central values and also expressing a basic emotional attitude. A number of different applications of gene technology have been used in the study as examples, which the informants have discussed and evaluated. The interviews also tried to grasp the informants' general knowledge of genetics and the source of that knowledge.

The interviews were carried out during the second half of 1996. The persons interviewed were found in the fields of social workers, farmers, biology teachers and nurses. The search for informants started with the apprehension that the general knowledge of genetics is rather limited in Sweden today. We wanted to include informants whom we anticipated as having some basic knowledge of the area. We therefore decided to include farmers and biology teachers, representing two groups that we expected to know a bit more than the average population about genetics. The nurses were included because of their closeness to medical care and also because they represent a group without higher university education. Social workers constituted an interesting group because of their social perspective upon human beings and society. In this way the groups represent different levels of education, occupational fields and knowledge of genetics.

Each interview lasted about one hour and was followed by a short questionnaire, which the informants filled out after the interview. Sex, age, nationality and living in town or in the countryside, as well as occupation, guided the subjective selection of informants. The questionnaire includes the questions designed to capture 'Sense of Coherence' (SOC) developed by Aron Antonovsky (1987). This is one attempt in the study to try to grasp the basic mood experienced by the informants.

The interviews started in the specific field of genetics and then continued over in a wider worldview discussion. This approach was able to open up and reveal unnoticed relations between this field of genetics and a broader interpretation of worldviews. Biological knowledge, and in this case genetics, touches upon many central apprehensions of the view of human beings, worldview and central values such as health and well-being. Through this connection between genetics and worldviews, the accent on genetics in the study was able to highlight and elucidate some worldview components.

One important question will guide the interpretations of the material. *Could there be any relation between worldview perceptions and attitudes towards gene technology, and if so, what do they look like?* One specific area that will be discussed in this paper is the ways the informants interpret nature. We were curious about what people mean when they say that the application of gene technology affects the natural order, or simply that it is unnatural. It seemed important to try to understand more about how one can interpret and understand nature, and the sense of order in nature that people are referring to. Is it simply an idea about certain limits that must not be overstepped? Or is it a conception of an all-embracing network, where everything and everyone is mutually dependent on one another?

The analysis of material has just started, and the remarks that will be made here will necessarily be preliminary and tentative. A few examples from the interviews will be presented, concerning some areas that at this stage stand out as interesting. We will start with some remarks on central values and interpretations of life in relation to gene technology and then say something about the ways the informants look upon nature. After that we will give an example of how Christian faith may relate to these topics of genetics and finally illustrate how attitudes towards gene technology may change, when related to specific concrete situations.

Values and Interpretations of Life

The most prominent values expressed by almost everyone interviewed, are the great significance given to family life, close relations and health. These areas all contribute to what some informants stress as personal well-being. These are patterns that we recognize from previous studies, on popular values and worldviews. In line with this, all the informants but one have no problem in accepting gene technology intended to cure severe diseases in man. Being able to cure diseases and diminish suffering among human beings is seen as positive and supports the central values expressed. Gene technology can in this way, by being put to clinical use, support these central values of health and well-being.

The one person who was very doubtful as to whether gene technology should be used for curing diseases is a female social worker. She has a personal orientation towards psychoanalysis and was herself some years ago close to losing her life, suffering from breast cancer. Her arguments are interesting. The main argument is based on the assumption that every crisis – even bodily diseases – leads to personal development and growth. She says;

> 'I don't think that a life that doesn't contain crises is meaningful. The crises create a possibility for me to make contact with the deeper part of myself and with the meaning of life.'

This is a concrete experience in her own life and she also recognize this in other people. For instance she sees Dostoevsky's illness as a source of creativity, and states 'that I would by no means have wanted to miss Dostoevsky'. This argument about crisis and human growth is supported by another argument. She question this new genetic technique and its implications concerning the way we look upon ourselves and upon death. As she puts it, 'Because there is a risk, as I see it, that you will reinforce the illusion that we are immortal.' In the entire interview this woman expresses the need for human beings to try to accept our conditions and learn to understand more about ourselves and life, and not try to escape or reshape the true human life. This position forms a background to her restrictive attitude towards gene technology.

Interpreting Nature

What about nature then? As we have seen in previous studies, Swedes have a very close relation to nature. The persons interviewed often express this warm feeling for nature and also suggest an inherent order in nature. This order is sometimes defined as the ecological system or the laws of nature, but for some informants this order may also include some kind of religious feeling. This feeling could be accompanied by a conception of a life spirit or by the view of nature as holy.

The course of events in nature is often described as something taking place without any interference from human beings. A crucial point in the ways people understand nature is to what extent, if any, we as *human beings are to be regarded as part of this order in nature.* Some informants express a very Darwinian understanding of the conditions in nature, and they are in line with the evolutionary theory, including human beings as just another animal that affects and is affected by the rest of the nature. There are other informants who point to the fact that we as human beings have in many ways destroyed and polluted our beautiful nature. This is a result of our effort to try to control nature, and because of this misuse we are not to be regarded as part of nature. According to this understanding, we would do best to leave nature to flourish on its own, and interfere as little as possible. There is also another important difference in the conceptions of nature that concerns the *values assigned to humans, animals and plants.* This is a difficult question to sort out, because many of the informants express a duality concerning this aspect of nature and values. On the one hand, many of them express very strongly that humans should not exercise any power over nature and are not to be valued higher than any other living creature, but on the other hand, because of the fact that we are human beings, we tend to favour our own kind. There are also informants that accept the human being as the crown of creation.

What are the consequences of these different concepts of nature, when turning to gene technology? The picture is ambiguous, but in some cases there seems to be a connection between the conception of the human being as part of nature and a more permissive attitude towards gene technology. And there are also examples of the opposite position, where informants emphasize that we are not to be regarded as part of nature, and that we as human beings therefore should not manipulate genetic material, or in any other way affect or destroy the natural order. The other aspect of this order in nature, the values assigned to different inhabitants on earth, sometimes supports this picture. When seeing human beings as part of the natural order and at the same time assigning them a higher value, the application of gene technology is easier to accept. When animals and sometimes also plants are given the same values as human beings, then it's harder to maintain our right to interfere in nature or to use other species for our own good. A female biology teacher may sum up this last opinion: 'We can't put ourselves at the top of this pyramid. I think that all that is living must be treated with respect.' And when asked about using organs from pigs, she said, 'How does this affect the animal? The animal must not come to any harm.'

The conceptions of nature seem to incorporate both the view of the whole of existence as a mutually dependent network and the view of every species – whether animals or plants – as having its particular purpose and place. In this system there is a balance that must not be disturbed. This raises two questions: firstly, is the order in nature that people perceive changeable? Are the many forms of life in nature undergoing constant change and in this change is there also balance? And secondly, regardless of whether or not the order in nature is changing, how easily is the balance disturbed? What is required to disrupt the balance? A major intervention of any sort or just the normal influence of everyday action? The answers to these two questions could be brought together. A perception of nature as changeable is easier to combine with a view of nature as resistant. The ability to manage change is then both a ground for protection and an effect of the interplay that is going on in nature and between human beings and nature. The opposite position, perceiving nature in a relatively static manner is well suited with the view of nature as fragile. If you approve of the last scenario, then it is less likely that you will regard nature as being able to manage genetic change without being fundamentally altered. In this respect biological knowledge could make a difference. Some of the informants with a more advanced biological education acknowledge a constant change of ecological systems. And due to this knowledge, they apprehend a capacity in nature to manage this change. The *ability to change* and the *resistance* in nature, the *value of human beings* and their *place* in nature, are aspects of nature or the world as a whole that might in different ways be related to perceptions of and attitudes to gene technology.

Religious Perceptions

The question of religious worldviews and their relations towards this new biology seems to be rather complex. This will be illustrated with an example from the interviews. We find in the material two persons who describe themselves as Christians, both men. One graduated from a school of social science and the other has been trained as a biology teacher. Their attitudes to gene technology differ quite a lot. The social worker expresses a very restrictive and negative attitude to gene technology, while the other man is rather positive towards a lot of applications. One way of understanding this is to take into account their different ways of interpreting the Christian creation myth. The social worker very much stresses God as a supreme being who once created this wonderful earth according to his divine will. And in relation to this belief, we as human beings should not interfere with the divine will and order. He says, 'When dealing with this [gene technology], you interfere with nature, in a way that we as humans ought not to do.' And he states, 'We try to change what God has created. And then we start to experience ourselves almost like gods.'

The biology teacher finds an argument for genetic engineering in the conception of the total creation as fallen in original sin. Therefore all of creation is incomplete and there is nothing holy in nature itself. 'But it is not holy in itself in any way, that no… I was thinking, in line with the Bible, it [nature] is also cursed because of man.' Because of this, we are allowed with care and proper knowledge to change genetic material, and he even accepts germline therapy in particular cases. Genetic engineering is for this man a way to help other people. 'You can regard this as God has given us these possibilities. And then we must use them.' The reason why this knowledge is available right now is interpreted by this man as an eschatological sign. '… in the end of days, which I believe we are experiencing right now, then we will have explored God's creation'.

Both these men share the conception that man has a higher value and a responsibility for the rest of the creation, but nevertheless, when they evaluate gene technology, they end up in almost opposite positions. There are many other aspects of this case that might help to explain their different positions. One is their difference in knowledge of the field of genetics, another is their general evaluation of technology, with the social worker expressing a more negative attitude to all sorts of technology, compared to the biology teacher. Another area only mentioned at the beginning is the emotional attitude of security or anxiety, optimism or pessimism. This kind of basic mood might influence an individual's attitude to gene technology as well as the rest of a person's life.

Different Levels of Attitudes

The last example from the material will illustrate the importance of a concrete situation, for being able to come to a decision about the different applications of gene technology. Some informants have great difficulties when they are asked to evaluate the different applications of gene technology presented to them. They spontaneously create different situations where the circumstances they add or subtract are important for the evaluation of the case discussed. This tendency is particularly prominent in relation to applications to human beings. A female farmer discussing the transplantation of organs from pigs to human beings says, 'But prolonging life of human beings. I am not so sure about that. I could almost say that I am against that.' In her opinion we should care more about each other while we are still alive, and not try to live forever. But her general judgement of gene technology is interesting as a contrast to her own reasoning on a personal level:

> But if it will concern me and my own family, I must know what I am deciding about. And that's what I mean, then you might not be consistent, and in that situation I will probably say, do everything that you can. And because of that, it's good that it exists after all.

This woman, as we can see, is conscious of the fact that the circumstances in the concrete situation will in a profound way affect the judgement of the technique discussed. In the material we can find examples of similar ways of reasoning, often by women. This tension between different levels of attitudes are sometimes also present in the perceptions of the informants, without being aware of it themselves.

The analysis of material will continue. We have today only given some brief and tentative examples here of perceptions and concepts that might be important. There are probably other important concepts that form components in personal and collective worldviews, that shape and reshape perceptions of gene technology. The perceptions of nature and the order in nature that people experience seem to be one important area to examine further. Whether or not there exists a relation between the informants' basic mood, captured by the SOC questions, and attitudes towards gene technology is an other interesting question to explore. Within this same research project we are presently planning a general survey of a representative sample of a Swedish adult population on the relationship between interpretations of nature and attitudes to genetic technology. This survey will provide us with more extensive data and will assist us in the analysis of the interview material.

References

Adler Carlsson, Gunnar. 1990. *Lärobok för 90-talet: Om vår överlevnads villkor*. Stockholm: Prisma.

Antonovsky, Aron. 1987. *Unravelling the Mystery of Health*. San Francisco: Jossey-Bass.

Halman, Loek, and Thorleif Pettersson. 1995. Individualisering och värdefragmentering: Resultat från The European Value Study 1981–1990. In *I tider av uppbrott: Värderingar och värderingsförändringar i det moderna samhället*, ed. Jonas Anshelm. Stockholm: Symposion.

Jeffner, Anders. 1992. A New View of the World Emerging among Ordinary People. In *Christian Faith and Philosophical Theology: Essays in Honour of Vincent Brümmer*, ed. Gijdbert van den Brink, Luco J. van den Brom and Marcel Sarot. Kampen. Kok Pharos Publishing House.

Keller, Evelyn Fox. 1995. *Refiguring Life: Metaphors of Twentieth-Century Biology*. New York: Columbia University Press.

Human and Animal Gene Transfers
Images of (Non-)Integrity in Greece
Eleni Papagaroufali

Human Organs vs. 'Genes'

In Greece, systematic research in molecular biology and biotechnology constitutes a recent and rather limited practice. Nevertheless concepts such as 'gene', 'genome', 'genetic code', 'genetic information' and the like have taken a prominent place in the public discourse. This is due to information circulating through the mass media and referring to applications of biotechnology, such as reproductive technology, as well as genetic and transgenic engineering. In some cases this information has positive connotations: it projects the curative prospects of predictive genetic diagnosis; it praises the possibility to prolong human life through transplantation of body parts coming from genetically engineered animals; it stresses the instances of health improvement through embryo genetic research. In most cases, though, it conveys messages that are negative, sarcastic and threatening: cloning and genetic diagnosis, related to embryo preimplantation, are viewed as leading to loss of democratic values, to discrimination and racism. Also, transplantation of 'humanized' animal organs is interpreted as leading to the loss of human species identity. By the same token, the possibility of eating animals with humanized organs has been presented as a way to cannibalism.

This negative stance is shared by small-scale ecological and animal rights associations. These categories of Greek people reject the idea that Nature, especially human nature, can be collapsed into genetic data. They are against transgenic manipulations of plants and animals and refuse the idea underlying them, i.e. that Nature is incomplete and needs improvement. The Greek Orthodox Church is also against such practices: their 'neo-evolutionist' background is considered 'incompatible with the Greek Orthodox Christian spirit' (Karoussos 1987: 46); also reproductive techniques, such as IVF, are considered adulterous. As for Greek physicians, there are many who tend to have a rather negative stance as well. Apart from those – mainly gynaecologists – who have established profitable enterprises – sperm banks, IVF centres and fertility clinics – there are those who question the possibility of developing solid legal and ethical support for such practices (e.g. Domenicou 1991), and those who view transgenic manipulations of animals and humans as anti-scientific (i.e. violating natural laws), inhumane and speciesist (e.g. Charitakis 1992). Also physicians involved in organ transplantations are projecting artificial rather than transgenic body parts as the only scientifically feasible solution in the future (Mandros and Kordatos 1991: 58).

In this article I will talk about a very specific category of Greeks and their ambiguous, if not negative, perceptions of genes and gene technology: it concerns adult men and women who have signed the so-called 'donor card', that is, individuals willing to donate after death and anonymously their organs to be transplanted to terminally ill persons, or their body to be dissected for medical education or research. Most of them are also blood donors.[1] Given the high degree of generosity of these people, one might expect them to be able to share any part of their body with anyone at any time, i.e., before/after death. This is not the case. Not all body parts are perceived or sensed as being of the same nature or value; not all of them are felt as having the same relationship to one's own self and the others. To begin with, prospective donors view human organs as 'parts of the organism', like legs, hands, blood. According to them, one may donate or receive organs to be transplanted 'without losing one's own personality or contact with the others', for these are mere 'things to be recycled' or 'spare parts'.[2] However, they do not share the same feelings about body parts that relate to genes, such as sperms, eggs, foetus cells. These are felt as identical with one's own self and are not available for donation nor accepted for reception. 'With organs, it's all in the bag', said one woman organ donor, in her early thirties:

> I give you or you give me an organ and it's finished, whereas with eggs ... you donate your genetic code... It's like donating a whole person ... you donate an entire human being... No, I wouldn't...

The same distinction between organs and gametes was made by a male organ donor in his mid thirties, father of two children:

> Organs and other body parts get incorporated into another organism which doesn't change in any way... It is a mere prosthesis... whereas the sperm is genes and creates a new life *in absentia*... You can't check upon the kind of crossing that will take place... I, for instance, would be worried that my sperm might be used to fertilize an orangutan or I would be worried that the child born might live a bad life...

Blood (donation or reception) is also distinguished from gametes (donation or reception):

> Sperm is different [from blood]. No, I would not [donate my sperm] because of the responsibility involved... Because I would transmit half of me of my genetic traits, and create descendants... It may sound irrational but from a sentimental point of view... I would like to know where it goes... With blood donation you cover a need. Sperm is not a matter of need but of concern and love...
> *(Male organ and blood donor in his late twenties)*

Implicit in these and other donors' statements is a distinction made between organs or blood and sperm or eggs, both in terms of their nature and the kinds of relationships imagined to develop in each case between donors and recipients. Thus, organs and blood are perceived as 'organic' entities, that is, as 'material', sensible, bounded, thereby manageable substances.[3] Sperm and eggs, on the other hand, are identified with 'genes', or genetic 'codes' or 'traits', that is, with entities of a monorganic, non-material, non-sensible, unbounded almost non-human nature.[4] Yet the fact that, unlike sensible organs/ blood, non-sensible 'codes' may produce very sensible and sensitive products, i.e. 'descendants', makes donors sceptical. Because in such cases what is going to be transmitted and shared (in life or after death) is not 'parts' of one's own 'organism' but one's own 'half self' (or 'half personality' or 'half identity'), the donors' concern about how it/they will be 'treated' by unknown recipients is high. The unknown destination and destiny of these non-material, yet potentially fleshy, entities causes donors both a fear that they will not be treated properly, i.e. with love, and a feeling of loss – loss of contact with one's own half self appropriated to unknown others, including descendants. By the same token, in cases of reception of sperm or eggs, the unknown origin of others' 'halves' causes puzzlement and anxiety. The following statement belongs to a woman body donor explaining why she would not accept a donor's sperm:

> This child would have half genes of me and half of this man... while I would be able to recognize 'me' in the child's personality I would worry about recognizing 'him'... I wouldn't know how to deal with this other half.[5]

For all these reasons prospective donors under study would neither donate nor receive 'genes'. Only a few said they would do it, provided they knew the 'identity' of the recipient or the donor: s/he should be a close friend who should love children, should love her/ his partner, and should accept that this transmittance takes place through physical contact (in the case of sperm). If these criteria of a more sensory and sensuous nature are lacking, then assisted procreation through donated gametes is considered 'too technological', 'too goal-oriented', a 'mere biological construction', 'abnormal', 'ugly', 'monster-producing', a 'mere combination of genetic traits', 'inhumane', 'non-human'.

In fact, all prospective donors (men and women) were against reproductive technology, in particular IVF, for 'if nature doesn't give you children, let it be so, let it go' – to use a young woman's words. In case of diagnosed infertility (their own or their partner's), they said they would adopt a child rather than receive a stranger's sperm or egg. The following statement by a woman donor is indicative of this stance:

> No, I wouldn't try anything like this because I wouldn't know where this sperm would come from… A criminal? A sick person? Instead I would adopt a child. You see it's like with transplantation. You adopt organs, you fulfil your needs and yet each entity keeps its own place intact

Humans vs. 'Genes'

Given the plethora of Greeks who nowadays either give their sperm – mostly for compensation – or receive sperm and eggs – in cases of infertility – what is it that makes these otherwise generous people so cautious (if not mean) with respect to the 'treatment' or the integrity/intactness of their 'genetic code'? What are they afraid of, when it comes to genes and gene technology, as opposed to technology related to organ transplantation and so on?

It may sound contradictory, but prospective organ or body donors are people extremely concerned about their bodily/fleshly selves and the integrity of their bodily/fleshly identity. They are appalled at the thought of their bodies being slowly decomposed by worms in the grave, and view donation (after death) as a way out of this self-dissolution. Although their bodies will disintegrate through organ extraction or dissection, and will decompose sooner or later in the grave, the fact that they are the ones who decide upon what will be done to their bodies upon death, makes them feel masters of their bodily selves and identity even after death. Practices related to the anonymous organ/body donation are not perceived as disrupting one's own identity and its imagined integrity.[6] As a body donor said, 'The funeral will take place on my picture, not *me*'. And according to an organ donor, 'Worms will not find much food to eat. *I* will have escaped them.' As for the anonymous organ recep-

Transgressing boundaries between humans and animals has always frightened people, as for example vaccination with the cow-pox virus (J. Gillray 1802).

tion, we are told that 'nothing changes' because it is only a matter of 'prosthesis' and or 'adoption'.

In other words, prospective donors are possessive and like to have the last word on themselves – and the others[7] – even *in absentia.* They like to think of themselves as semi-creators and full controllers of their historically and culturally specific presence in life and after death. Yet when 'genes' come into their body picture, these imagined advantages disappear. Anonymous donation of one's gametes or 'codes' is seen as producing anonymous creators and controllers of one's own identity, as well as uncontrollable creations – humans blurred with or dispersed into animals? On the other hand, anonymous reception of others' gametes – other than one's own partner's – is seen as producing (semi-)anonymous and uncontrollable creations. Unlike organs, strangers' genes are not perceived as entities that can be simply 'adopted' so that the integrity of the original codes or individual identities may be kept 'intact'.[8] On the whole, gene technology seems to be too uncontrollable a practice for people so possessive of themselves – and others.

The same bodily image of integrity and fear of its loss (through gene blurrings and dispersion) were found in the donors' stance towards receiving certain kinds of organs (Papagaroufali 1996a). Donors were asked whether, in case they found themselves in a state of terminal illness, they would consent to become recipients of either human, or animal organs, including genetically altered ones, or artificial organs, in order to prolong their lives. No one responded negatively. Most donors preferred artificial vs. animate (human or animal) organs. Next came those who would prefer human and artificial vs. animal organs, and least of all were those who would prefer animate (human or animal) vs. artificial ones.

All preferences depended on what was perceived as more 'normal' or 'natural', meaning in this case closer to human nature, so that the original identity stays as intact/integral as possible. Thus, for instance, the high degree of 'naturalness' attributed to artificial organs was based on the fact that they – like eye-glasses and contact lenses, dental seals, pacemakers etc. – are made/constructed by humans. Cadaveric (vs. living) organs, organs coming from baboons (vs. pets) and organs stemming from living strangers (vs. friends, relatives) were seen as less 'natural' and were therefore less preferred. Yet the kind of organs that were perceived as least or not at all natural – and hence were totally rejected – were the ones that would come from genetically engineered animals – i.e. transgenic pigs. No one – including those donors who would have preferred only animate (i.e. human or animal) organs – was attracted by the idea that these organs would be 'humanized'. On the contrary, they considered them either 'monstrous' or 'too artificial'. In fact, artificial organs, compared to transgenic ones, were seen as more 'human'(-ized), i.e. as 'having more of a soul, metaphorically speaking' because they have 'a direct connection with human hands and mind'; by contrast, in trangenic organisms (and organs) it was said it is 'genes' that 'do the job' through the 'mediation' rather than 'direct' intervention of humans. The donors' fear that their own 'genetic code', and hence both their individual and human-species identity, might be lost or blurred with, or controlled by these 'things' inside them, surfaced once again.

What is important in this case – as in the case with donation/

reception of gametes – is that donors seem to perceive 'genes' (including their own) as something separate or different from 'humans' (including themselves), and express more mistrust in them than in those who carry or manipulate them. In general, donors' stance towards 'genes' (associated with sperms, or eggs, or transgenic organs) seems ambiguous if not negative: Although they endow them with personifying characteristics, such as intentional action (they are the ones who 'do the job' or create 'descendants'), they perceive them as less-than-human (soul-less? less 'tangible'? less 'mind'-ful?) and simultaneously more independent and powerful than humans.[9] Although they consider genes worthy of their 'love' and 'care' – as opposed to using them 'artificially' to cover 'mere biological needs' – at the same time they want to control them. Although they know their own descendants, through egg or sperm reception, will have at least 'half' of their own or their partner's genes, they prefer to adopt a child than let this 'half' mingle with the unknown 'other half'. Although they know that mingling of the two 'halves' occurs also in the case of procreating with beloved partners, they prefer to think that 'knowing' the other means keeping each one's 'half' (and the halves incorporated into the descendent) intact, that is, controlling it/them.

History and Society vs. 'Genes'

It is obvious that although the donors' definition of nature (or natural-ness) is broad, it includes entities that are sensed either as the same as, or familiar to themselves (e.g. animal organs should at least come from pets, and genes should at least come from or go to close friends). Underlying this 'xenophobic' stance there is the (Western) idea that there is an 'original' nature (human and other) that should be preserved intact/integral and should be reproduced.[10] The fear that this primordial nature might be damaged or lost is not unique to specific donors nor, as was previously said, to the Greek Church and Greek activists. It is also shared by citizens of countries where bio and gene technology blossom. Thus there are European and American activists who are against many kinds of genetic research – including the Human Genome Diversity Project (e.g. Balsamo 1995; The 'Vampire Project' 1995; Haraway 1995). Also, there are lawyers and environmental policy makers who are against most transgenic manipulations. They are worried about the possibility of transgenic organisms being released 'in the wild' and altering the 'original' state of the ecosystem (McCarthy et al. 1994: 28). A similar stance is also taken even by full supporters of such practices. Even molecular biologists, who consider the notion of species integrity a scientific heuristic device with taxonomic (vs. biological) significance, suggest that different contexts of biotechnological practices – biomedical laboratories, markets, farms, and 'in the wild' – should be kept separate and each one should entail separate moral obligations (e.g. Singleton 1994). What all these people imply, including the specific prospective donors to a more extreme extent, is that the world of the new techniques, of new 'data' and of newly 'coded' creatures should be kept somehow separate from the older one. Apparently the latter is perceived by many people (if not by molecular biologists and geneticists) as more 'original', (vs. 'new'), more 'organic' (vs. 'code'-based), more 'natural' (vs. 'engineered'), more 'normal' (vs. 'monsters'), more integral/intact (vs. 'trans'-genic), thereby more human and real.

What is interesting, however, with respect to the specific donors, is that, despite their rather negative stance towards this new world, they are already embedded in its discourse and imagery. Indeed, in the context of the specific discussions, they were readier to depict their bodily selves and make claims to the integrity of their human identities in 'genetic' terms than in the ones associated with the 'organic' world. Thus, as was shown, sperm and eggs were identified with 'genetic codes' rather than with parts of the organism. Also, personal and descendants' identities were claimed on the basis of genetic traits or relations rather than 'blood' relations. Finally, worries about the preservation/reproduction of the(ir) human species were expressed at the level of the future manipulation or engineering of genes rather than in terms of origin stories of a religious character (creationist approach) or of a natural evolutionary one (descent from chimpanzees etc.). Put otherwise, although donors tend to associate the 'genetic' body image with situations tending to various sorts of hybridity, they unconsciously use it to defend the integrity/intactness of their personal and human identity. But the 'unconscious' is 'the forgetting of history which history produces by incorporating the objective structures it produces in the second natures of habitus', says Bourdieu (1977: 78).

It seems that the historically and culturally specific practice of gene technology is already in the process of becoming 'second nature' or 'habitus', that is, history forgotten. This phenomenon is perfectly exemplified in an advertisement by Levi's blue jeans in which the picture of a woman wearing jeans is accompanied by the pompous statement – in English: 'These genes made America'! Underlying this play on words is the developing and domineering idea that the 'genetic' rather than socio-historical makeup of someone or of whole nations is the locus of both their creation and evolution or progress – and, later on, of their 'second genesis'.[11] Put otherwise, it is 'genes' rather than historical human beings who create societies – such as 'America'! – and make them progress. Anonymous, contextless, a-historical genes appear as powerful entities, (a kind of 'mana'?), who 'do the job' while historically and culturally specific, organic or flesh-ly humans only 'mediate' in it – to use the donors' words.

Apparently there are many people, including Greeks, who have already adopted this ideology and related practices: e.g. many men and women determined to have children of 'their own' (rather than adopted) do not seem to worry about where the received eggs and or sperm might come from as long as these 'do the job' effectively. Also, many terminally ill persons will definitely accept 'humanized' animal organs, when these finally become available on the market. Prospective donors, though embedded in this discourse or ideology, were shown to resist its practices. Yet the main reason underlying this resistance is the same as the one underlying its acceptance: the reproduction and maintenance of, and the control over the 'original' sacred image of the same, or the familiar, who is the Human Self – to use Haraway's words.

If this interpretation is correct, then, we might suggest that both positive and negative (public) perceptions of genes and gene technology are but the contemporary version of origin stories so much cherished by Westerners, both those who have the power to 'engineer' the world to become 'better' than in its 'original' state, and those who have no other choice but resist, or succumb to this 'engineering'.

Notes

1 This chapter constitutes part of a wider project on human organ and body donation after death, in Greece. It is based on interviews and long discussions with 24 urban, middle-class, well-educated men and woman (Athenians), aged from the mid-twenties to mid-forties. Most of them are body donors (8 women and 6 men); the rest are organ donors (6 women and 4 men) (Papagaroufali 1996a).

2 Despite these characterizations, not one donor would sell (or buy) organs or other body parts because 'the value of one's own body cannot be assessed with money'.

3 According to some donors, blood donation is 'more real' than organ donation: In one case, the (living) donor can 'see', even 'touch', the 'gift' (blood), and 'knows that blood goes where it is needed'. In the second case, the (dead) donor cannot 'see' or 'feel' the whole procedure, and 'cannot know whether the organ goes where it is needed'.

4 Haraway has noted that practices related to gene technology produce entities (or images of entities) of a 'different ontological kind than flesh-and-blood organisms, "natural races", or any other sort of 'normal' organic being'. Put otherwise, the 'human' aspect of genes/genomes is equated with, or represented as ontologically specific things called 'information', or 'codes', or 'database' that can be 'circulated', 'decoded', 're-designed' (1995: 352). Apparently, donors have adopted this less-than-human/non-fleshly image of genes and, therefore, distinguish them both from flesh-and-blood human organism and from flesh-and-blood humans (see later in the text).

5 During our discussions, women tended to perceive themselves more as recipients (of sperm) than as donors (of eggs). By contrast, men tended to see themselves more as donors than as indirect sperm recipients – i.e. through their female partners. Perhaps, as a consequence of these identifications, men were slightly more concerned about the destination (and destiny) of (their) human genome, whereas women were slightly more concerned about the destiny of Nature – due to biotechnological interventions.

6 Donors share the same feelings with respect to cremation: the 'immediate' disappearance of one's own body, as opposed to its gradual decomposition, is perceived as 'immortalizing' one's own self the way it is while in life or immediately before death, i.e. integral/intact.

7 This possessiveness extends to relatives and other 'own' people: donors would not donate the body or organs of own people upon their death, because they would like to keep them the way they used to be when alive, i.e. intact/integral (Papagaroufali 1996c)

8 In 'normal' cases of conception where partners 'know' and or love each other, there is the feeling that neither 'blurring' nor 'dispersal' of hereditary traits such as genes (as opposed to blood) takes place, so that the offspring is a brand 'new individual'

(on this see also Strathern 1992: 53) while parents keep themselves 'intact'. By contrast, when conception is based on gametes of others than one or both partners, then there are some (e.g. ethicists, lawyers), who consider it essential to know who the biological antecedents 'really' are: blurred and dispersed traits must be disentangled and reamassed respectively, to identify the various kinds of 'realities' required by society – e.g. genetic vs. birthing mother, genetic vs. legal father etc.

9   Emily Martin (1991: 500), drawing also on R. Petchesky (1987), shows how biotechnology 'pushes [us] back indefinitely' to tinier and tinier entities (e.g. foetuses or sperm/egg at the cellular level) and makes us perceive them as (inner) 'persons', independent of the (outer) persons who carry them, and entitled to more legal rights than the latter.

10 On this, see Haraway's excellent analyses, 1989, 1991

11 Rabinow has coined the term 'biosociality' to describe the process by which 'sociality' is now re-found through a re-making of nature-as-culture, i.e. of nature as artificial technique. The example he uses is the Human Genome Project, that is, the plan to rewrite the 'Book of Man' and to produce a 'second genesis' by reproducing heredity and evolution as artificial technique – rather than as natural fact (cited in Franklin 1996: 14; see also Rabinow 1995: 449–51). The process here is not one of modelling culture or society on nature (e.g. sociobiology), but one of modelling culture on nature-as-culture or technique.

References

Balsamo, A. 1995. Forms of Technological Embodiment: Reading the Body in Contemporary Culture. *Body and Society* 1 (3–4): 206–215.

Charitakis, G. 1992. *Doctors' Silence*. Athens: Lygouras and Co.

Domenikou, A. 1991. Transplants From Animals. *Ethnos,* 19 November.

Franklin, S. 1996. Science as Culture, Cultures as Science. Paper presented at the Fourth EASA Conference, Barcelona, 12–15 July, 1996.

Haraway, D. 1989. *Primate Visions: Gender, Race and Nature in the World of Modern Science*. New York and London: Routledge.

Haraway, D. 1991. *Simians, Cyborgs, and Women: The Reinvention of Nature*. New York: Routledge.

Haraway, D. 1995. Universal Donors in a Vampire Culture: It's All in the Family: Biological Kinship Categories in the Twentieth-Century United States. In *Uncommon Ground:*

Toward Reinventing Nature, ed. William Cronon. New York: W. W. Norton.

Karoussos, K. 1987. *The Man From the Monkey? An Answer to the Materialist Approach*. Athens: Chrysopigi Publications.

McCarthy, C. R., and G. Ellis. 1994. Developing Policies and Regulations for Animal Biotechnology and the Protection of the Environment. *Hastings Center Report* 24, 1: 24–29.

Mandros, S., and D. Kordatos. 1991. Transplantations Today. *Iatriko Vima* 1991: 47–61.

Martin, E. 1991. The Egg and the Sperm: How Science has Constructed a Romance Based on Male and Female Roles. *Signs: A Journal of Women, Culture and Society* 16 (3): 485–501.

Papagaroufali, E. 1996a. Xenotransplantation and Transgenesis: Im-moral Stories about Human–Animal Relations in the West. In *Nature and Society: Anthropological Perspectives*, ed. P. Descola and G. Pálsson, pp. 240–55. London and New York: Routledge.

Papagaroufali, E. 1996b. Body and Organ Donation after Death: Multisensory Perceptions of the Body. Paper presented at the Conference 'Contemporary Trends in Anthropological Research: Identity, Body Representations and Material Culture', Panteion University, Athens, 25–26 April 1996.

Papagaroufali, E. 1996c. Body and Organ Donation after Death: Practices of Remembering and Forgetting. Paper presented at the Fourth EASA Conference, Barcelona, 12–15 July, 1996.

Rabinow, P. (1995) Reflections on Fieldwork in Alameda. In *Techno-Scientific Imaginaries: Conversations, Profiles, Memoirs*, ed. G. Marcus, pp. 155–176, 449–451. Chicago: The University of Chicago Press.

Singleton et al. 1994. Transgenic Organisms, Science and Society. *Hastings Center Report* 24(1): 4–14.

Strathern, M. 1992. *After Nature: English Kinship in the Late Twentieth Century*. Cambridge: Cambridge University Press.

The 'Vampire Project'. 1995. *Hastings Center Report* 25(1):6.

Consumer Acceptance of Genetically Modified Food

L. J. Frewer, C. Howard and R. Shepherd

Introduction

It has long been recognized that public reaction is likely to be a crucial factor in developing and introducing biotechnology (Cantley 1987: 5–15; De Flines 1987: 6–9; Stenholm and Waggoner 1992: 22–35). Various factors have been identified as important to the development of public resistance to technologies. In particular the media (Sparks, Shepherd and Frewer 1995: 267–285; Marlier 1992: 52–108) and the impact of the legislative framework by which the technology is regulated (Jasanoff 1995: 335–356) have been recognized as influential.

It is difficult to dissociate ethical issues from those of perceived risk and benefit (Straughan and Reiss 1996), and it is essential that these concerns be addressed within the context of both regulation and release of GMOs, and strategic development of the technology. Novel or new foods resulting from biotechnological treatments raise questions relating to conceptual, ethical and technological issues, as well as toxicological and nutritional elements associated with particular products (Knudsen and Ovesen 1994: 365–369). It is also apparent that differences in perceptions are likely to exist both between individuals and between cultures. Questions arise within the context of regulation and development regarding these differences in attitudes. Equity of distribution of both risks and benefits is unlikely to accrue equally across all population groups, and this is likely to further influence attitudes and beliefs about the technology (Foreman 1990: 121–126).

The strategic development of genetic engineering in food production is likely to be market-led, rather than science-driven, as ultimately it is the consumer who will decide whether to purchase food products in the supermarket. It is notable that issues of consumer perception and consumer acceptance were not addressed during the development of food irradiation. The potential of the technology was not realized, because of consumer resistance to irradiated foods.

Early Assumptions about Public Perceptions of Genetic Engineering

Early research into questions surrounding public perceptions of genetic engineering in food production made two assumptions which now appear to be erroneous.

The first was that public attitudes towards the development and application of genetic engineering would be driven by beliefs intrinsic to the science *per se*, rather than how it was applied. Unlike the food irradiation example given above, genetic engineering cannot be regarded as a 'unitary' technology. Different applications are likely to be associated with different beliefs about risk and benefit. For example, medical applications are likely to be more acceptable than food-related applications, on the grounds of both perceived risk and ethical concern (Frewer and Shepherd 1995: 48–57). Within an application area (for example, food production), beliefs are likely to further dissociate according to process considerations (for example, the type of organism involved in the transfer, the benefit of applying the technology, and to whom) (Frewer, Howard and Shepherd 1996a: 61–67; Frewer, Howard and Shepherd 1997b).

The second assumption was that the public could be 'educated' to accept the technology, and that simply providing information about the underpinning science would convince the public that the technology was acceptable. There is evidence that increasing public understanding may further polarize views – those who are in favour of genetic engineering may become more positive in terms of their attitudes, those who are negative may become more focused in terms of their opposition (Evans and Durant 1995: 57–74).

The Social Construction of Risk

It has long been recognized that risk perception is socially constructed (Fischhoff, Slovic, Lichtenstein, Read and Combs 1978: 127–152). Factors other than technical risk estimates guide the behaviours and actions of individuals regarding particular hazards – the extent to which a hazard is perceived to be new, involuntary, and uncontrollable represents one dimension, the second is determined by the extent to which the hazard is likely to result in fatality and in terms of its catastrophic potential. Genetic engineering is typically perceived as a hazard of moderate severity and one which is extremely unknown in terms of the consequences should the hazard occur (Sparks and Shepherd 1994: 79–86).

Expert views of risk assessments do not take account of this viewpoint that risk is 'socially constructed', and disparities between the expert and lay perceptions are frequently observed for different hazards (Sjöberg and Drottz-Sjöberg 1994). In particular, there is evidence that scientific experts are more positive towards the development and application of the 'new genetics' (Michie, Drake, Bowbrow and Marteau 1995: 243–255). However, it is the psychological constructs which determine risk perceptions which also determine public reactions and behaviours to potential hazards. It is essential that the psychological characteristics which determine risk perception be understood in order that the public be informed not only about risks and benefits of applying genetic engineering, but are also provided with information about their other concerns.

In order to understand what these characteristics and perceptions actually are, it is essential that qualitative methodology be applied in order to elicit actual and realistic concerns, rather than to use researcher-generated rating scales to assess what psychological factors are associated with public resistance to a particular technology. One method employed to address this problem is to use semi-structured interviewing in conjunction with analytical techniques such as generalized Procrustes analysis (Kelly 1955). The aim is to determine what are the real concerns of the public regarding different food related risks (Frewer, Howard and Shepherd 1997a: 98–124).

In two separate experiments, the underlying concerns respondents expressed regarding different applications of genetic engineering drawn from food-related, agricultural and medical areas were elicited. In the first study, applications were phrased in very general terms. In the second study, a different group of people was asked to respond to more specific applications, where the tangible benefits of the technology were more obvious. Twenty-five different respondents were included in each group. Both sets of data were submitted to generalized Procrustes analysis, and graphical representations of concern about applications generated. From this, the combined results of both studies were used to design a survey instrument (n=200). Survey research and the application of principal components analysis indicated that respondents saw applications of genetic engineering which were presented in general terms as either positive (necessary, beneficial or advantageous) or negative (risky, unethical, harmful or unnatural), whereas the specific applications tended to be more highly differentiated in perceptual terms, such that negative perceptions linked to some applications were mediated by perceptions of need or benefit. Some applications were seen as low in risk, but also unnecessary. Replication of this research using an Italian sample produced an almost identical pattern of rejection or acceptance for the different applications – although in the Italian sample, concerns did not focus on risk. Instead, respondents expressed concern about the ethical issues associated with the different applications of the technology (Saba, Moles and Frewer submitted). This effect was attributed to the lower level of media coverage about genetic engineering in Italy relative to the UK. The results indicate that public concerns are not simply driven by perceptions of risk. Failure to ask appropriate questions may result in erroneous predictions about public acceptance of genetic engineering.

L. J. FREWER, C. HOWARD AND R. SHEPHERD

Effective Communication about Genetic Engineering

The goal of effective communication about any potentially hazardous technology is to ensure that members of the public are provided with the best scientific information available regarding the potential risks and benefits of applying the technology. In the case of genetic engineering in food production, such an information strategy will enable the consumer to make an informed choice about consuming products of the technology, as well as contributing to the wider public debate about strategic development of genetic engineering.

The failure of the nuclear industry and authorities to gain public support for continuing developments has been considered to result from the counter-productive efforts of the public relations industry, rather than a systematic attempt to develop effective risk-benefit communication about nuclear radiation risks (Greenhalgh 1994). Issues of trust in the risk regulators is also likely to be important, particularly if they are the source of the information provided.

Petty and Cacioppo (1984: 668–672) have extensively studied the area of persuasion and the impact of information on attitude change. The elaboration likelihood model represents one of the most important theories of persuasive communication which has evolved within social psychological theory. The basic premise of the model is that there are two routes to persuasion, the 'central' route and the 'peripheral' route. The central route involves in-depth processing of the incoming information, whereas the peripheral route utilizes external cues surrounding the information to permit simple inferences about the merits of the content without recourse to complex cognitive processing. There is considerable support within the social psychological literature to support the model (Eagly and Chaiken 1993). Whilst its utility in the area of health-behaviour change (Fife-Schaw 1989: 16–21) and risk communication (Frewer, Howard and Shepherd 1996b: 48–52) has been identified, there is also potential for applying the model to the development of effective risk-benefit communication strategy. It is likely that factors such as trust in the information source may represent an important social cue in terms of how the information is interpreted (Frewer, Howard, Hedderely and Shepherd 1996: 473–486).

The application of the Elaboration Likelihood Model has demonstrated that source credibility is an important determinant of the reactions of an individual to information about genetic engineering. The Elaboration Likelihood model was used to systematically investigate the impact of source factors (credibility and perceived risk immediacy of information) and informational factors (persuasive content of the information) on attitudes towards genetic engineering, and whether these factors resulted in greater thought about the technology. The results from 160 respondents indicated an interaction between source credibility and persuasive content. People tended to respond more to information low in persuasive content if it was attributed to a high credibility source, and to persuasive information if it were from a low credibility source. However, under these circumstances, responses tended to express concern about the technology. It was concluded that source credibility is an important determinant of public responses to information (Frewer, Howard, Hedderely and Shepherd submitted).

It is also important that consideration of prior attitudes be included in the development of effective communication. Social Judgement Theory is a communication model developed within social psychology which takes account of the potential importance of prior attitudes in determining responses to incoming information. Essentially, the model adopts an attributional approach which features mechanisms of attitude formation and change which can occur in the absence of argument-based processing. A substantive review of the theory is provided elsewhere (Eagly and Chaiken 1993). However, attribution approaches emphasize how peoples' inferences about the cause of communicators' attitudinal statements affect their agreement with these statements. The primary emphasis in Social Judgement Theory is on the importance that prior attitudes have on reactions to incoming persuasive information. Research originating within Social Judgement Theory has indicated that the negative impact of low source credibility can be mediated by admission of risk uncertainty, but that extreme attitudes are difficult to change through interventions, at least for those with extreme attitudes (Frewer, Howard and Shepherd, in press).

Future Research

Given current knowledge about genetic engineering and public perceptions, various questions arise regarding public understanding of the technology, and how these impact on perceptions. It is essential that effective intervention studies for public understanding of genetic engineering be developed, and for the impact of public understanding of science on attitudes and acceptance issues systematically identified. Some individuals may prefer to maintain 'scientific ignorance', whilst others may actively seek out scientific information about risks and benefits of technological application. If this is the case, then research might usefully identify how and why individuals react differently. Furthermore, there is clearly a need to examine the long-term impact of information interventions on risk perceptions, and beliefs about benefit resulting from genetic engineering. Finally, public attitudes are unlikely to remain stable, particularly as the products of the technology are made available in the market-place. Further research into the relationship between crystallizing public opinion, and behavioural choices regarding the consumption of genetically modified foods, will inform both attitudinal theory and policy development.

Acknowledgements

Parts of the work reported here were funded by the UK Ministry of Agriculture, Fisheries and Food and by the UK Biotechnology and Biological Sciences Research Council.

References

Cantley, M. 1987. Democracy and Biotechnology: Popular Attitudes, Information, Trust and Public Interest. *Swiss Biotech* 5.

De Flines, J. 1987. Publieke opinie essentieel voor biotechnologie, Interview. *De Ingenieur* 6.

Eagly, A. H., and S. Chaiken. 1993. *The Psychology of Attitudes.* New York: Harcourt Brace Jovanovich.

Evans. G., and J. Durant. 1995. The Relationship between

Knowledge and Attitudes in Public Understanding of Science in Britain. *Public Understanding of Science* 4.

Fife-Schaw, C. 1989. The Elaboration Likelihood Model of Petty and Cacioppo: Implications for HIV/AIDS Education Campaigns. *Health Psychology Update: The British Psychological Society* 4.

Fischhoff, B., P. Slovic, S. Lichtenstein, S. Read and B. Combs. 1978. How Safe is Safe Enough? A Psychometric Study of Attitudes towards Technological Risks and Benefits. *Policy Sciences* 9.

Foreman, C. T. 1990. Food Safety and Quality for the Consumer: Policies and Communication. In *Animal Biotechnology in the 1990's: Opportunities and Challenges*, ed. J. Fessenden MacDonald. Ithaca, N.Y.: National Agricultural Biotechnology Council.

Frewer, L. J., C. Howard, D. Hedderely and R. Shepherd. 1996. What Determines Trust in Information about Food-related Risks? Underlying Psychological Constructs. *Risk Analysis* 16.

Frewer, L. J., C. Howard, D. Hedderely and R. Shepherd. Submitted. Reactions to Information about Genetic Engineering: Impact of Source Credibility, Perceived Risk Immediacy and Persuasive Content.

Frewer, L. J., C. Howard and R. Shepherd. 1996a. The Influence of Realistic Product Exposure on Attitudes towards Genetic Engineering of Foodstuffs. *Food Quality and Preference* 7.

Frewer, L. J., C. Howard and R. Shepherd. 1996b. Effective Communication about Genetic Engineering and Food. *British Food Journal* 98.

Frewer, L. J., C. Howard and R. Shepherd. 1997a. Public Concerns about General and Specific Applications of Genetic Engineering: Risk, Benefit and Ethics. *Science, Technology and Human Values* 22.

Frewer, L. J., C. Howard and R. Shepherd. 1997b. Understanding of Risk Perceptions Associated with Different Food-Processing Technologies – A Case Study Using Conjoint Analysis. *Food Quality and Preference* 8:271–290.

Frewer, L. J., C. Howard and R. Shepherd. In press. The

Importance of Initial Attitudes on Responses to Communication about Genetic Engineering in Food Production, Agriculture and Human Values.

Frewer, L. J., and R. Shepherd, R. 1995. Ethical Concerns and Risk Perceptions Associated with Different Applications of Genetic Engineering: Interrelationships with the Perceived Need for Regulation of the Technology. *Agriculture and Human Values* 12.

Greenhalgh, G. 1994. *Experiencing Disaster: Risk and Impending Disaster in a Historical and Social Perspective.* Stockholm: Stockholm School of Economics.

Jasanoff, S. 1995. Product, Process, or Programme: Three Cultures and the Regulation of Biotechnology. In: *Resistance to New Technology,* ed. M. Bauer. Cambridge: Cambridge University Press

Kelly, G. A. 1955. *The Psychology of Personal Constructs: A Theory of Personality.* New York: Norton.

Knudsen, I., and L. Ovesen. 1994. Assessment of Novel Foods: A Call for a New and Broader GRAS Concept. *Regulatory Toxicology and Pharmacology* 21.

Marlier, E. 1992. Eurobarometer 35.1. In *Biotechnology in Public: A Review of Recent Research,* ed. J. Durant. London: Science Museum.

Michie, S., H. Drake, M. Bowbrow and T. Marteau. 1995. A Comparison of Public and Professionals' Attitudes towards Genetic Developments. *Public Understanding of Science* 4.

Petty, R. E., and J. T. Cacioppo. 1984. Source Factors and the Elaboration Likelihood Model of Persuasion. *Advances in Consumer Research* 11.

Saba, A., A. Moles and L. J. Frewer. Submitted. Public Concerns about General and Specific Applications of Genetic Engineering: A Comparative Study between the UK and Italy.

Sjöberg, L., and B. M. Drottz-Sjöberg. 1994. *Risk Perception Of Nuclear Waste: Experts and the Public.* Report No. 16. Stockholm: Stockholm School of Economics.

Sparks, P., and R. Shepherd. 1994. Public Perceptions of the Hazards Associated with Food Production and Food

Consumption: An Empirical Study. *Risk Analysis* 14.

Sparks, P., R. Shepherd and L. J. Frewer. 1995. Assessing and Structuring Attitudes towards the Use of Gene Technology in Food Production: The Role of Perceived Ethical Obligation. *Journal of Basic and Applied Social Psychology.*

Stenholm, C. W., and D. B. Waggoner. 1992. Public Policy. In *Animal Biotechnology in the 1990s: Opportunities and Challenges,* ed. J. Fessenden MacDonald. Ithaca, N.Y.: National Agricultural Biotechnology Council.

Straughan, R., and M. Reiss. 1996. *Improving Nature? The Science and Ethics of Genetic Engineering.* Cambridge: Cambridge University Press.

The Social Production of Nature
Gene Technology, the Body, and New Conflicts
Alberto Melucci

Where is Nature?

In today's complex societies Nature is increasingly a culturally interpreted reality, a reality in which we intervene ever more directly. More than in any society of the past, scientific knowledge and technology intervene in our natural environment and our biological structure. We have developed the reflexive capacity to 'produce' our own reproduction and the environment itself. This increased intervention in our inner Nature and the Nature surrounding us takes the form of a pure symbolic capacity allowed by science, as today testified by gene technology. In societies which depended much more closely on Nature, human action was made manifest in its material products. Now that mankind's capacity for action on the natural environment and its own Nature has grown so great that we have acquired the power to actually produce or destroy it, our capacity for action has become relatively independent of its products and is being transformed into a pure reflexive capacity which acts upon its own natural roots.

The laws of Nature will not operate as the 'pure' natural forces they once were: they may still turn against us if ignored or violated, but they will become part of our everyday experience, of our personal and collective choices, of our moral concerns. No other culture of the past has been exposed to the necessity of deciding of its own limits in a comparable manner for the simple reason alone that no previous society has developed a capacity to 'produce' itself to such a degree and with an unlimited power of self-destruction.

In everyday life and at the global scale the definition of biological 'normality' and 'pathology' becomes a contested issue: individual needs for autonomy come into conflict with a standardizing rationality of technical apparatuses, which use scientific knowledge for intervention in the deep-lying roots of human behaviour and its environment.

Achieving an equilibrium with 'Nature' can thus only be the outcome of a surplus of awareness and of an explicit cultural choice. It can never be produced regressively by an ingenuous belief in Nature as a 'good Mother' whose deep laws are spontaneously transmitted to their children. The difference between our culture and all those of the past is that we now possess the power to deeply intervene in nature, including our own inner nature. As a consequence, natural laws no longer exist independently of society; the romantic dream that it would still be possible to recreate a pure nature which can deliver us from the necessary evils of civilization is an illusion or an ideological mystification. This fact notwithstanding, there still exists a Nature which can be culturally safeguarded and respected by recognition of it and by responsibility for it. The paradox and the challenge is that we may, culturally, become Nature. This, moreover, is in fact our only chance in a world which by now is almost completely constructed by ourselves – a world whose 'natural' dimension we can either bring into existence or destroy.

However, we are still factually born, we fall ill and die; we are more than ever exposed to the extremely delicate constraints of the ecosystem in which we live. But the difference between us and the cultures of the past is that these restrictions become an object of cultural awareness and are culturally elaborated far beyond the simple recognition of 'natural laws'. Other cultures have simply recognized and accepted such constraints as imposed on them by the authority of a higher being or a given superior order. Now it is a matter of our own decision whether we are to survive or suffer a catastrophe, whether we are willing to accept, reject, or overcome our limitations. It is not enough to throw ourselves upon the whims of nature; we have to choose and 'create' nature (starting from respect for it) through a productive expansion of culture and of awareness.

Colonizing or Inhabiting Life

Granting that the questions related to gene technology signal these transformations, one must also admit that the ways in which modern science has historically constructed its relationships with the environment, with other species, and with non-industrial cultures reflect the limits of our awareness, and the violence we have committed on that portion of ourselves that resist instrumental rationality. Through an increasing intervention on our mental, psychological, and biological life, today we have started creating the 'inner planet' (as opposed to the external planet as our natural environment). The inner planet is not a natural datum, but nor is it a magical domain where dark forces operate. We take action to redraw its boundaries and map out its topography, we take possession of its territories, trying to subject the new frontiers to the jurisdiction of rational control and to the transforming action of purposeful projects.

Today, new conflicts arise as we move in to appropriate this inner planet. On the one hand, we have learned how to participate in the formation of our identity, to take conscious action on ourselves, to explore and occupy the lands of our interior. On the other hand, we are denied this opportunity by the increasingly invasive intervention of the apparatuses of control and regulation, which with ever-advancing capacity aim to define the coordinates of the inner planet, set arbitrary borders, and lay claim to our motivations, affects, and our biological structure itself.

As today's individuals, we possess the ability and the option to consciously intervene in the production of our capacity for action; we can adjust our motivations, come into a closer contact with our bodies, work on our emotions. At the same time, however, we remain subject to processes of external manipulation which, in the name of techno-scientific rationality, colonize the inner planet and control the motivational, affective, and biological roots of our behaviour.

On the one hand, then, our potential for autonomous action, our reflexive ability to produce meaning and motivation for what we do, increases. We have available to us a potential for individuation – an opportunity to inhabit the territories of the inner planet as individuals; that is, to become individuals in the fullest possible sense – to an extent unknown to any previous epoch in the history of the human species. On the other hand, however, we are exposed to a parallel increase in the powers of control over the formation and transformation of our identities, to an erosion of the margins of our individual independence, and to an intensifying social regulation of our behaviour that tacitly forces us to manipulate our most intimate dimensions, even the biological ones.

In education, in the definition of health and sickness, of normality and pathology, our need for autonomy comes into conflict with a standardizing rationality which uses scientific knowledge for intervention in motivational structure, for the pharmacological regulation of behaviour, for the orthopaedic manipulation of interpersonal relationships that sets the standards of the 'correct' behaviour in having sex, raising children, making friends, forming couples; and now the standards are increasingly set also for our biological 'normality'. Occupying ourselves with the inner planet therefore also directs our attention to the ways in which it may become a land of conquest and a tutelage of external authority or a space for autonomy and responsibility.

On the other hand, the deterministic patterns we inherited from positivism have dissipated and a circular conception of the relationship between biological life, mental experience, and social reality is assuming their place. Holistic models have been constructed which are at odds with the reductionist tradition of modern science and which affirm the impossibility of separating microcosm and macrocosm. The inner planet is no longer an essence, but an articulation of levels and systems which alters the way we perceive ourselves. The dualistic heritage of the mind-body relationship has been called into question, and we have begun to leave the dominant model of linear causality behind. The body as a relational vehicle also expresses our inner nature and translates the program drawn up by our biological structure into behaviour. The body, however, is not a machine commanded by the mind; rather, it 'embodies' the mind and enables us to exist as unified wholes.

In safeguarding and developing the inner planet we have to prepare to fight against its colonizers and we can incorporate this enormous wealth of knowledge into our field of experience. We must learn to explore, to settle, and to cultivate rather than passively submit. The lands of the inner planet are limitless and we have only just begun to explore them; the next few years seem destined to reveal further unknown continents. And even at this moment of the exploration, new conquistadores are equipping themselves to set off in search of the golden continent represented by the biomedical business. It should, however, be possible for every one of us to become both an explorer and a custodian of these lands, so unique and so intimately part of us. Taking charge of them, protecting them against the dangers that threaten, cultivating them with respect and without violence, are duties which we are already called upon to fulfil. It is a task which involves each and every one of us, but which we nevertheless cannot treat as an individual problem. The emergence of corporeal, cognitive, and relational processes as a field of deliberate intervention signifies the transformation of behaviour into messages, into meaningful discourse – that is, into a discourse capable of conveying the deep-lying link between the within and the without, between the nature that we are and the nature that we inhabit.

Co-living and Responsibility

If we accept that not everything in our relationship to nature is entirely calculable and that not everything can be accounted for by instrumental rationality that still guides science and technology, we need an ethic that does not insulate us against the risk of choice and which enables us to metacommunicate about the goals of choice and the criteria that underlie the decision. Such an ethic emerges as irreducibly situational, as an ethic that preserves the dignity of the individual decision and forges anew the bonds that bind us to the species, to living beings, to the cosmos. Stretched between the two poles which mark the confines of the human condition, removed from nature by our capacity for language and restored to it through our bodies, we must look to language and to the body as the foundations of an ethic responding to the need to cope with the problems of a planet by now wholly shaped by our intervening action.

The body maps the confines of nature within and without us: the great rhythms of birth and death, the permanent cycle of day and night, of the seasons, of growth and ageing. In a world that technology has constructed, equilibrium, rhythm, and respect for limits are no longer – if they ever were – the spontaneous outcome of a nature as 'mother and mentor'. Rather, they are the fruits of individual and collective choices, of a conscious morality which both assumes *responsibility for* and *responds to* nature.

It is precisely this dual significance of responsibility that roots ethics in language. Culture is the realm in which every moral choice must necessarily take shape. In our planetary society of information, to name is to bring into existence. The simplistic idea that information mirrors a 'reality in itself' is a residue from the past whose authority we must renounce. Information *is* reality – if only because our experience is by now entirely *mediated* by the representations and the images that we produce. Nature, as the foundation that grounds the experience and as the external reality that grounded the question of truth and falsehood has ceased to exist independently of the effects of social intervention. Culture becomes the space where reality will be defined and Nature itself preserved or destroyed.

What, then, is to be done with language – this is the new frontier of an ethics of complexity. How and for what purpose should we use the *power of naming* which allows us to fabricate the world and to subsume it to the signs with which we express (or do not express) it? Power does not only lie outside us as a threat to be exorcized. If instead of projecting it externally we realize that it is a relation, then it becomes constitutive of all our relationships while still remaining in our own possession. It is language which takes up the challenge of creation of meaning or its reduction to signs. Through empty or meaningful words, words which hide or which reveal, it is within language that nature can be violated or brought to existence.

63

New Conflicts

We are witnessing a denaturalization of nature and culturalization of conflicts. We are moving from the idea that there is a nature and a body out there, as natural sources of physical and biological events, to the awareness that nature and the body are entirely defined within culture. The consequence is that conflicts will increasingly affect the definition of the same field: not just what to do with it but how to name it. Once there is no nature outside the domain of social action intervening deeply in it, the problem will be the agreement or disagreement on the definition of what 'nature' is: and this is not just a matter of words, but implies thoroughgoing 'material' consequences in technology, economy, organizational power. Nonetheless, the conflict is played out at the level of definition, because the 'experts' increasingly proclaim to us what 'nature' is in a given field (biogenetics, sexuality, medical issues) and the people affected by this intervention will refuse to understand nature the way those experts want to define it. The field of foreseeable conflicts will then be increasingly within culture, and not between society and something conceived as an external, supposedly pure nature.

We are already facing these types of debates in human reproduction or the genetic issues, where the knowledge is always incomplete and often unclear, but entails according to the definition produced by institutional apparatuses significant practical consequences on the policies, the allocation of economic resources, and other arrangements affecting the common life.

Discourses and practices related to the ends of social life are inherently conflictual: there are always two sides to the same issue. The social discourse on gene technology exemplifies this profound ambivalence. What is important in this respect is not the discourse itself, but the diffusion of practices concerning the body and personal needs (from sexuality to reproduction, to health issues) that have changed people's everyday lives. In none of these practices is the body entirely reduced to discourse. Culture shifts its attention to the body, but it can never entirely frame the body and render it into a mere message or a symbol: there is always a part of bodily experience which is not translated into language. Deep feelings, emotions, sensations, and movements are not entirely communicable to others because they also represent the deepest and most intimate parts of personal experience.

This is why information works, simultaneously and in an ambivalent way, in two opposite directions. On one hand it improves individual capacities to make conscious choices, to be aware of what one experiences in one's own body, to prevent diseases and enhance one's physical and psychological well being. On the other hand, however, information creates new uncertainties: knowing more about the body, its functioning, its biological and physiological structure, broadens the field of available options, both in terms of interpretation of body signals and in terms of decision making. In which scientific category should one put a physical message coming from the body, or in which explanatory frame provided by the available scientific language should one interpret the signs (sensorial, visual, kinaesthesic) that constitute the flow of our bodily experience? And moreover, what kind of decisions and actions should follow from one's interpretation of those messages, given the fact that we know more about the possible answers and solutions available? Knowing what science knows and how it can intervene presents us with a range of possibilities that imply choice and decision making, therefore uncertainty.

But, even more dramatically, scientific information diffused and incorporated in people's everyday life creates a cognitive screen that might reduce the capacity of directly perceiving one's own body, sensing it, appreciating its feelings: instead of experiencing the living body one superimposes the cognitive filter of scientific categories transforming the bodily experience into medical diagnosis, anatomical topography, or physiological mapping.

There are many examples of this ambivalent role of information: genetic and sexual education for children and adolescents, scientific books for children on human body, physiology, reproduction, television educational information, medical education and preventive information for adults, sexology manuals and courses, all operate as informational channels as well as cognitive and normative filters organizing the perception and relation to the body. The implicit normative prescriptions on the 'correct' behaviour one should adopt as far as the body is concerned are certainly one of the most striking effects, often unintended, of scientific information.

Another important example refers to the increased diagnostic and preventive capacity that gene technology is able to provide. Genetic deficits, inherited or genetically transmittable diseases can be now traced and diagnosed through a sophisticated set of available instruments that are increasingly incorporated as routines practices in preventive health policies. While allowing targeted and effective intervention, these new possibilities also create new problems. A pregnancy test revealing a genetic problem exposes the subjects involved (pregnant women, foetuses, family) to new and previously unknown problems, fears, sometimes dramatic choices. A diagnosis confirming the genetic transmission of a disease with painful and lethal development (often already known through the experience of other members of the family) can add to the real and foreseeable suffering the desperate anticipation of that very suffering. These are completely new problems created by the power of knowing: they certainly cannot be dealt with only within the logic of medical apparatuses and techniques and need to be placed in a new ethical, cultural, and social framework.

Information creates new problems also if one considers the social use of the available knowledge. Information can easily become a new and sophisticated instrument of social control. The main example today is offered by the increasing pressures made by insurance companies in order to get the most accurate and trustworthy information on individual health conditions. Because money is directly involved and coverage depends on the nature of information available, we should expect increasing efforts to control the criteria and the techniques for medical diagnoses. This trend is already apparent in the field of mental health where the logic imposed by insurance companies increasingly follows this linear sequence: nosological definition of the trouble; diagnosis; standard timing for treatment according to the diagnosis. The real condition of the person, how the person experiences his/her suffering, what are the personal needs connected to the illness become less and less relevant and the external definition prevails imposing its pre-established procedures.

Finally, our culture still relies on a notion of information that is entirely cognitive as if the fact of 'knowing' were sufficient to convey and contain the meaning, in matters that are deeply intimate and very personal. The affective implications of information, the ambiguities, and ambivalences related to our affects are rarely considered and often left aside in medical practices and institutional policies concerning personal needs.

The shift towards a bodilization of our culture is therefore neither a process of 'liberation' nor just a new form of rationalization and of hidden manipulation. It contains the seeds of a deep contradiction, for bodily experience inalienably belongs to the individual and only individuals can 'practice' the body. Once the process of an experiential approach to the body is triggered through new forms of awareness, it can never be entirely controlled again. If culture allows people to consciously experience their body, there emerges at least a part of what is lived by individuals which escapes the social discourse, is not included in it, and cannot be entirely controlled. There develops therefore a contradictory potential for change which has been introduced through the diffusion of body practices. Alternative medicines, for example, have rapidly fuelled a new market of professionals, institutions, and health products, but the very fact that through these practices people have an opportunity to experience a different relationship to their body keeps the construction of meaning at least partly open and prevents a complete commodification and manipulation of the body.

We can imagine a society where the capacity for manipulating the body will increase: a deeper penetration into the structure of our genes and into the chemistry of emotions is not just science fiction anymore. Yet, unless we think of a system of total domination, there always remains a part of our bodily experience which cannot be entirely culturalized, which belongs to individuals and can become a nucleus of resistance or opposition against external manipulation. This rootedness of human experience in the body may provide a potential for change in overculturalized social systems.

Putting the Boundaries

Complex systems rely on a new faith: the faith in science. Nevertheless when facing his/her present pain and suffering the person is not helped in any manner by the fact that science will solve the problem in the future. But at the collective level and when one is not directly affected, the faith in science is an effective way of dealing with uncertainty. Since we do not know enough, we are not competent, we rely on science and we trust experts. The motto 'In God We Trust' could today be substituted by 'In Institutions We Trust': we rely on institutions based on expertise and on the fundamental faith in science.

This new faith in science and expertise creates a new paradox: in complex systems, we are facing an overmentalization of human faculties, even a transferring of human capacities in machines which will contribute to a further expansion of the cortical power of the human species. This overcerebralization of the collective brain, now increasingly embedded in our technology, undermines all the other human capacities related to our feelings, emotions, movement, biological rhythms. The other human 'brains' which connect our species to the evolution of life on Earth ('brains' which govern our biological and sensory experience) are simply denied or ignored. But these seemingly archaic levels of experience, which link us to the chain of biological evolution but are also part of human culture, return and surface in different forms, in illness, everyday ailments, weakening of immune systems, and the like. Instead of being simply denied, these symptomatic signals can be heard and interpreted, and they can be incorporated as part of human experience. The body can be either 'normalized' through external intervention (medication, intervention on brain chemistry, manipulation of biological rhythms) or become the field for an autonomous and meaningful experience of the many potentials of human species, far beyond its cortical power. This different attitude towards the body is not just an individual choice which can change the direction of everyday life; more than that, it can be the deep texture of broader cultural changes.

This is why we will face new conflicts related to the body. Attitudes towards gene technologies show that one-third of the population is in favour. The supporters are educated, male, urban, and correspond to the stereotype of the yuppies. One-third is against and the opponents are divided between traditionalists who express the fear of disruption of the old order (older, less educated, and rural) and the 'greens' (young, educated, urban, female) who oppose these technologies because of the risks they imply. The new social conflicts are likely to oppose the first and the third group, who share the same cultural field but interpret it in opposite ways: they both belong to and highly differentiated, highly technological society based on information but they struggle for different values and orientations.

The social regulation of genetic interventions and the management of these conflicts deeply challenge modern democracy and can only take place through an increasing visibility of power. Power should be made visible at different levels (individual, as in doctor-patient relationship; institutional, as in diagnostic or preventive procedures; public, as in the definition of policies). The necessity of putting a limit which is not pre-established but must be set through some form of decision-making creates an inevitable degree of selection and exclusion and a quota of shadow and silence. Therefore it is critically important to have arenas where these voices can be heard and become public concerns (for example, consensus conferences). The level at which we decide to set the boundaries of 'Nature' in gene technology will be a fundamentally cultural, political, and social decision. There is nothing pre-established, nothing a priori; there are only the decisions as to where to put the limit, knowing that in every case it must be drawn somewhere because we are not entirely cultural beings, we always remain partly natural beings, we live in an environment, and we are a nature in ourselves. But where biology and nature end and culture begins is not decided outside culture. This is the new paradox. We are thus overcultural beings facing the necessity to decide on our own nature. This, again, creates an enormous issue of responsibility with which human beings have never had to cope before and for which we need a new solidarity, because this cannot be just an individual endeavour. We need to concentrate our collective energies to face our fears and weaknesses: instead of relying on our power (scientific and technological) alone, we should start sharing our pain, the deep fears that always accompany big leaps in culture. A consciousness leap is required to face this passage of an era and it cannot be an individual or psychological experience: it must be a social and collective task. I wish these pages could contribute to this task.

References

Arditti, Rita, Renate Duelli Klein, and Shelley Minden. 1984. *Test-Tube Women: What Future for Motherhood?* London: Pandora Press.

Bonnicksen, Andrea L. 1989. *In Vitro Fertilization: Building Policies from Laboratories to Legislatures.* New York: Columbia University Press.

Corea, Gena. 1985. *The Mother Machine: Reproductive Technologies from Artificial Insemination to Artificial Wombs.* New York: Harper and Row.

Featherstone, Michael, Mike Hepworth, and Bryan S. Turner (eds.). 1991. *The Body: Social Process and Cultural Theory.* London: Sage.

Falk, Pasi (ed.). 1994. *The Consuming Body.* London: Sage.

Field, Martha. 1988. *Surrogate Motherhood: The Legal and Human Issues.* Cambridge MA: Harvard University Press.

Lundin, Susanne, and Lynn Åkesson (eds.). 1996. *Bodytime: On the Interaction of Body, Identity, and Society.* Lund: Lund University Press.

Melucci, Alberto. 1996. *The Playing Self: Person and Meaning in the Planetary Society.* Cambridge: Cambridge University Press.

Melucci, Alberto. 1996. *Challenging Codes: Collective Action in the Information Age.* Cambridge: Cambridge University Press.

Merchant, Carolyn. 1990. *The Death of Nature: Woman, Ecology and the Scientific Revolution.* San Francisco: Harper.

Strathern, Marilyn (ed.) 1993. *Technologies of Procreation: Kinship in the Age of Assisted Procreation.* Manchester: Manchester University Press.

Visions of the Body

Susanne Lundin

'Come along on Lennart Nilsson's new voyage through inner space.' The world media urged us with these words to watch the 1996 winner of the prestigious Emmy award for best television documentary. Over thirty years previously, Lennart Nilsson had amazed the world with his photographs showing how a human being develops from an embryo into a newborn baby. Since then he has penetrated deeper and deeper into the microcosm, showing us pictures of white corpuscles attacking invading bacteria, or the beautifully purple-coloured but lethal AIDS virus, or chromosomes enlarged to fantastic proportions. And now, most recently, the prize-wining voyage of discovery through the evolution of various animal species during the foetal stage.

Scientists through the ages have been engaged in similar exposures of the mysteries of the body, fascinated with charting our innermost landscape. Leonardo da Vinci's fifteenth-century pictures of nerve-paths and blood vessels, and his depiction of the womb are astounding illustrations of this curiosity. Yet it was not really until the nineteenth-century Darwinian revolution and the rapid growth in the numbers of doctors that biology and the human organism became a specific research area. In pace with the development of modern technology, more and more entrances to the very interior of life have been opened up. Virtually every tiny part of the body and its genetic structure has been scrutinized, analysed, and named.

This knowledge process is also a precondition for modern medicine, which is fondly described in terms of scientific objectivity. Yet the eye that examines these pictures of the body is pre-programmed; its vision is selective and culturally determined.[1] New versions of biological truths are thus being created all the time, new definitions of what a natural body is and what social functions are associated with it. My aim here is to investigate this interplay between the interpretation of scientific facts and people's personal attitudes to this knowledge; to discuss how modern optics not only make the innermost parts of the body visible but can also become a meeting place for both old and new ideas about the body. In this article I direct the gaze inwards to study how biological processes are depicted, described, and internalized in human consciousness.

Pandora's Box

'Once there was a tiny cell, smaller than a grain of sand. Although that cell was very small it carried an incredibly complicated and amazingly clever plan... ...A plan to make a unique living creature... YOU! How did that plan fit inside one tiny cell? How was it able to put all the parts of you together in the right order? Well, the plan to make you was coiled up in a tangle of very thin threads inside the first cell. To learn about those threads is to learn about the secret of life.'[2]

This is the opening of an illustrative and colourful children's book which both my children have devoured enthusiastically. My six-year-old daughter shuddered with delight when she heard that there was a secret deep inside her body, for when the cell began to divide, the chromosomes appeared, and they contained all the information needed to create a unique human being. It was a bit horrific, my daughter thought, to imagine that there was something so important inside the body that she didn't know about. My twelve-year-old son, on the other hand, a keen Sherlock Holmes reader, felt challenged to solve the fascinating mystery. He quickly became a real gene detective.

I retell this episode as an example of how an advanced science, such as gene technology, reaches into our everyday lives. Genetics and genetic engineering do not just belong in laboratories and hospitals. They enter every nook and cranny of everyday life – all the way into the nursery. This knowledge of complex scientific phenomena, and the way biomedicine and gene technology are integrated into people's everyday lives, is the subject of the following discussion (cf. Strathern 1992). It is a question of the norms and values that emerge in our biotechnological society. But it is also about how people's self-understanding and identity are shaped on the basis of the knowledge that it actually is possible to explore our bodies, right down to the smallest chromosome.

Today's technology gives the potential to stretch the limits of the body: for example, when a man suffering from a hereditary disease such as cystic fibrosis, which can lead to infertility, becomes a father thanks to sperm donation; or when reproduction technology is used in a completely different way, by which human DNA is inserted into pig embryos which have been fertilized *in vitro* – this way of cultivating 'human' organs may be used for patients who might need a kidney transplant.[3]

Such medical treatments presuppose a method that makes it possible to examine the body in the smallest possible detail. Fibre optics, microscopes, ultrasound, foetal diagnostics, and genetic screening are some of the ways of looking into the body. These technological tools can penetrate our bodies and also dissolve, not only the boundaries between humans, but also between different species.[4] It is thus the body that is the focus of my research. Or rather: body, technique, and identity. It feels fruitful to start with the body because it is significant capital for modern people, perhaps precisely because, thanks to various technologies, there is such potential in the body (cf. Turner 1996; Melucci 1996; Frykman 1994). It is like Pandora's box – a horn of plenty containing both good and bad.

Biology as a Cultural Filter

The children's book that I started with, *DNA is Here to Stay*, is part of a small series in which the other titles are *Cells Are Us*, *Cell Wars*, and *Amazing Schemes within your Genes*. They all illustrate very clearly today's view of the body. Or rather, they give instructions

Figure 1. Children are encouraged to learn about the genetic code, to reveal the biological map (*DNA is Here to Stay*).

about how we should relate to our bodies; how we should use biology as a controlling mechanism in existence. Biology has become the filter through which we are expected to interpret the world.

I mentioned that my son was transformed into a gene detective, determined to solve the DNA mystery. The first page of *DNA is Here to Stay* shows a tiny cell hiding a secret deep inside and saying: 'Guess what?' At the end of the same book young readers can glimpse a DNA data bank (Figure 1). Here they get to know that it is possible and important to see through the body, to learn about every single detail and to crack the genetic code. Children are thus encouraged to follow clues that can reveal the biological map, the plan of life that is hidden in their own body. The message is clear: we humans are constructed according to specific genetic templates. It is almost like a jigsaw puzzle – it is just a matter of putting the right piece in the right place.

I enjoy browsing through my son's children's magazine, *Kamratposten*. It too lets us clearly see today's view of body and nature. The magazine has a column about medicine and psychology, where readers can learn everything about such topics as how sperm production works, or about sweat hormones. One articles discusses menstruation. The text is illustrated with a colourful picture of a smiling girl showing a reproductive factory with an ovulation centre and tiny gear-wheels. Other texts deals with what happens when the body system is out of order, for example why some people get allergic when patting cats. 'Something is wrong in the complicated immune system', says the article. A funny picture of a child's body, constructed like a castle where a battle is going on between white blood cells and histamine, shows how we can win this battle by medication (Figure 2). In this magazine there are also letters to the editor. A girl writes: 'I'm a 12-year-old girl and very scared right now. I can feel a lump in my vagina, do I have cancer?' She receives the comforting answer that it is the visible part of the cervix, the portio, which has an important function for her as a female.[5]

Thus the body is presented as a complex system concealing both good and bad, but also as a machine; one has to know how to use the apparatus to be able to control it. In these circumstances it is not surprising that the young girl sees her interior filled with frightening things, that she connects strange body parts with complex diseases.

Practically everywhere we look, we find this biological-functionalistic imagery – partly similar to the sixteenth century's conceptions of the world and body as mechanical systems (cf. Merchant 1980; Sachs 1996). Yet these ideas do not exist just in the children's world, as we have seen, but also in the everyday media flow of instructive explanations, as when xenotransplantions are made clear with flow charts (Figure 3). A frequently used device is to let the reader follow the transgenetic process through three pedagogical steps. Step one: human genes are injected into a pig's ovum; step two: breeding of transgenetic pigs; step three: transgenetic organs are transplanted into human beings (*The Independent*, 1 May 1996). The media also discuss, for example, the importance of sunlight for our psychological well-being, as when showing a head in section pointing out where the hormone melatonin is situated and how it is activated through darkness. The headline 'January without sun activates the hormones of the dark' illustrates the seemingly simple association between biology and behaviour (cf. Oudshoorn 1994).

Multiplicity Images

It is not just in books and the media that biological awareness – a kind of biological gaze – is communicated and expressed. In numerous other contexts today, we get an opportunity to direct our gaze inwards. In the medical arena especially, people become involved in the interpretation of biological phenomena.

Through the lens of the microscope we can see, for example, sperm swimming round. We can find out which of them function well and which of them are sluggish. In my study of *in vitro* fertilization I have conducted interviews with childless couples. A male informant receiving treatment for his low sperm count told me once about when he was urged to have a look in the microscope. The doctor wanted the man himself to see how slowly the sperm moved. 'That wasn't so nice of course,' he told me, 'to see those poor lethargic things. To see such clear evidence of your own failure and unmanliness.' Another informant told me about a gynaecological examination, about how difficult it felt to have the doctor's hand probe deep inside her, while keeping her eyes glued to the ultrasound screen, where she could see her Graafian follicles floating like balloons, and how she felt an affinity with these organs.

Being subjected to the scrutiny of the medical eye means being transilluminated and described in scientific terms. But this process is not just one-way. It also gives a person an opportunity to see and name alien things. A woman who had suffered several extrauterine pregnancies was able to tell me exactly what had happened inside her body: 'Because the egg is in the fallopian tube, this sends a message to the pituitary gland and down to the womb to say "Look

Figure 2. The body is presented as a complex system containing both good and bad. But also as a machine: one has to know how to use the apparatus to be able to control it (*Kamratposten* 1995, no. 17).

out, here comes an egg." And so the womb gets ready. But the egg never comes because it has got stuck, and then the womb doesn't really know how to act. The egg control centre doesn't work.'

My informant describes the path of the egg through the body as if she had accompanied it on the journey, equipped with her camera. It is also as an observer and an explorer that she reports on her meeting with this other world. This woman has clear pictures of her interior. She conjures up a bodily machinery which has developed a fault that makes her uterus lose control. But her body is also a biological territory where cells and organs have taken on human features, making her pituitary gland and her womb seem like rational beings which unfortunately let themselves be deceived by an egg going astray.[6]

It is common for my informants to describe themselves in this way. They depict the body as an ingenious piece of apparatus which you can learn to master. The internal elements are important constituents, and all of them must function in harmony. At the same time, one may suspect that the personification of these organs and cells also conceals a darker, destructive feeling, the sense of being – for better or worse – at the mercy of a machine governed by biological rationality, which seems to be populated by purposeful actors: the ovum, the uterus, and the fallopian tubes.

There is a broad discussion indicating that we live in a hypervisualized time, in a society where the lens of technology has become the most important metaphor for the eye (cf. Petchesky 1987; Jay 1993; Ristilammi 1995). Many have stressed that the corporeality of the media leads to fragmentation and loss of identity. But it is not always the case that the interior of the body becomes something that the self finds it difficult to identify with and reconcile itself with. A while ago I met a woman who had finally become pregnant after many attempts at IVF. She said that, when she was connected to the CTG machine, which measures the contractions of the uterus and displays them as a curve on the screen, 'Although it felt a bit strange to *see* how you *feel* by looking at a screen, it was still something special to get this *proof* of my pregnancy.' For this woman, the contractions visualized by this technology, combined with her own palpable labour pains, were a confirmation of genuine motherhood (cf. Lundin 1996).

When people encounter the practice of medical care, this naturally has consequences for their own self-understanding. Technology gives rise to a visuality that brings the interior of the body to the outside; this may be an image on the ultrasound screen, the embryo in a test tube, or the image resulting from a genetic screening. Visible images are thus created of what has hitherto been concealed, but also linguistic images, metaphors, to describe this unknown world. Judging by symbols such as 'egg control centre' and 'wise cells', the new knowledge does not just bring alienation. Metaphors like these also formulate a will to make these organisms into something wholly personal and individual. This reminds us of earlier cosmologies where the body was interpreted as a homogeneous unity from which no part should be taken away. At the same time, there are close points of similarity between these ideas and today's wish to present the body as a personal project.[7] Obviously there is more than one possibility for the images of the interior body and the self-identity to coincide. Thus there are not only different interpretations of biological facts, but also numerous historically rooted images of the self.

Reflexivity and Responsibility

The revelation of our interior is a technological practice leading to cultural and individual reflexivity. It may sometimes mean that we think that we get to know our real selves. This is evident, for example, from the statement of the young pregnant woman seeing her female identity expressed by a machine. Or, in a different way, from the media information about melatonin, showing that our more depressive temperament in the dark of winter is a genuine biological sensation and not mere imagination, not a case of acquired laziness that has to be overcome. In a similar way, there is a lot of talk today about our genetically predetermined characteristics: dispositions that conjure up pictures of men as testosterone-packed males and of women as being born destined to care for others. Characteristics firmly anchored in the body give legitimacy to existing gender roles.[8]

At the same time, as we have seen, biological awareness may have the result that we become strangers to ourselves and what we bear inside us. One can speak about one's own organs, cells, and genes as if they were parts of a different universe. Or we can react as my

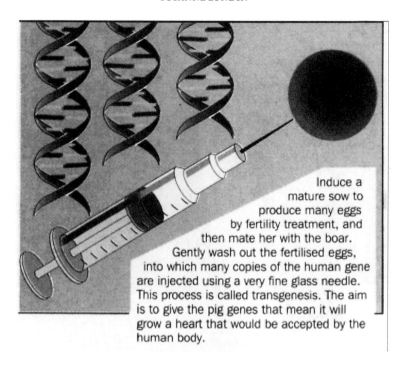

Induce a mature sow to produce many eggs by fertility treatment, and then mate her with the boar. Gently wash out the fertilised eggs, into which many copies of the human gene are injected using a very fine glass needle. This process is called transgenesis. The aim is to give the pig genes that mean it will grow a heart that would be accepted by the human body.

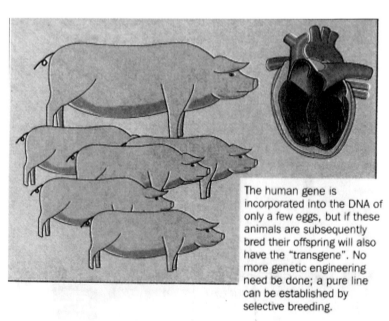

The human gene is incorporated into the DNA of only a few eggs, but if these animals are subsequently bred their offspring will also have the "transgene". No more genetic engineering need be done; a pure line can be established by selective breeding.

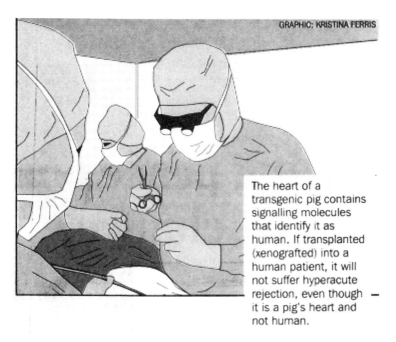

GRAPHIC: KRISTINA FERRIS

The heart of a transgenic pig contains signalling molecules that identify it as human. If transplanted (xenografted) into a human patient, it will not suffer hyperacute rejection, even though it is a pig's heart and not human.

Figure 3. Xenotransplantation is illustrated in biological-functionalistic imagery (*The Independent*, 1 May 1996).

six-year-old daughter did when she heard the book *DNA is Here to Stay*, shuddering at the thought that there is something deep inside her body, controlling her without her knowing anything about it.

Of course, these reactions need not be mutually exclusive. In general, however, one can say that the functionalist model of the body agrees well with today's cultural statement about the determinism of genetics – that we are at the mercy of our biology. Yet there is a normative commandment here. We are exhorted to remember that, even though the genetic order is all-powerful and unshakeable, it is still possible to influence it – especially when we humans learn how to crack the genetic code. In this way, genetics and genetic engineering acquire a kind of discursive effect (cf. Rapp 1993; Rose 1994). The cultural message is incontrovertible. The body is certainly a self-willed apparatus. But knowledge brings the potential to break down its structures and then to put them back together, hopefully in a much better state than before.

So we certainly cannot say to ourselves, 'Oh, I'm too tired to go to work today, my melatonin hormones want me to take sick leave,' or 'I don't want to have an amniocentesis. The child is welcome, with or without congenital defects.' No, our biological competence and our knowledge of the natural order – combined with the technical facilities on offer – mean that more or less explicit demands are made of us: we must assume responsibility. Not just by finding out about hereditary diseases for the sake of the family. The responsibility may also be borne on a more immediate level, for example, by controlling one's own reproduction with the aid of contraceptives. Or by carrying out a medical examination of one's own in the home, as the girl who wrote the letter to the children's magazine did, when she discovered a strange growth. Or when pregnant women check the albumin content in their urine by means of reagent strips that they can use in their own bathroom.

In other words, the signals are clear: we should take responsibility, and it is appropriate to begin with our own body.[9] The way people in their everyday lives are dealing with the new technologies shows that the body is not primarily understood as a functionalist and mechanical model, but rather as a flexible apparatus embedded in complex biological and social systems. As I have discussed above and elsewhere, people's self-identity is not constructed by one image of the self, but by a wide spectrum. Yet, being part of such complex systems is problematic. The consequence of thinking in terms of complexity, as the anthropologist Emily Martin points out, might be described as the paradox of feeling responsible for everything and powerless at the same time.[10] In this way both individual reflexivity and responsibility are accompanied by a kind of empowered powerlessness.

Notes

1   Martin (1991) describes science as a producer of world views, a knowledge that is hidden behind a mask of 'true facts'. See also Keller 1987; Rose 1994.
2   Quotation from *DNA is Here to Stay*, by Fran Balkwill and Mic Rolph, winners of the Science Book Prize for 1991.

3   In the project 'Transformations of the Body' I have devoted the last few years to a study of the general problems associated with *in vitro* fertilization (Frykman 1994; Lundin and Åkesson 1996; Lundin 1997; Åkesson 1997). Through the project 'Genethnology: Genetics, Genetic Engineering, and Everyday Ethics', this research field has recently been expanded to comprise a number of different empirical areas. At the centre there are a large number of diseases for which genetic analysis or genetic engineering prove useful.

4   Cf. Ingold 1994; Papagaroufali in this volume and 1996; Wheale and McNally 1995; see also Haraway 1991.

5   *Kamratposten* 1995 no. 17, 1996 nos. 15, 19. Cf. Martin's discussion (1994).

6   Lakoff and Johnson (1980) show how we create metaphors of our internal organs in which they are presented as human beings. The personification of organs is a way to transform these scarcely comprehensible parts of the body into something we can understand. See also Ideland 1997.

7   In every culture there are parallel thought systems which derive nourishment from traditional patterns coupled with contemporary expressions. See, for example, Lynn Åkesson's 'The Message of Dead Bodies' (1996), where she shows that the idea of the wholeness of the body has been significant in many historical periods, not least today; this is evident, for instance, in the reluctance of many people to donate organs.

8   R. W. Connell (1995) discusses the two opposing conceptions of the body that have dominated in recent decades. In one the body is a natural machine which produces gender difference – through genetic programming, hormonal difference, and so on. In the other approach the body is a more or less neutral surface on which a social symbolism is imprinted. See also Giddens 1993.

9   Our approach to the body has deep historical roots in earlier ideas of biology and nature. Unique for today is the normative aspect of individual choice of freedom and responsibility.

10 In *Flexible Bodies* Emily Martin (1994: 122) discusses how we learn to feel responsible for our own health, how we learn that personal habits such as eating directly affect our health. But we also learn that wider circles such as family relationships and work situations are related to personal health. She explains this cosmic view of the body: 'Once the process of linking a complex system to other complex systems begins, there is no reason, logically speaking, to stop'.

References

Åkesson, Lynn. 1996. The Message of Dead Bodies. In *Bodytime: On the Interaction of Body, Identity, and Society*, ed. Susanne Lundin and Lynn Åkesson. Lund: Lund University Press.

Åkesson, Lynn. 1997. *Mellan levande och döda: Föreställningar om kropp och ritual.* Stockholm: Natur och Kultur.

Balkwill, Frank, and Mic Rolph. 1992. *DNA is Here to Stay.* London: HarperCollins.

Connell, R. W. 1995. *Masculinities: Knowledge, Power and Social Change.* Berkeley: University of California Press.

Frykman, Jonas. 1994. On the Move: The Struggle for the Body in Sweden in the 1930s. In *The Senses Still: Perception and memory as material culture in modernity,* ed. C. Nadia Seremetakis. Boulder: Westview Press.

Giddens, Anthony. 1993. *The Transformation of Intimacy. Sexuality, Love and Eroticism in Modern Societies.* Cambridge: Polity Press.

Haraway, Donna J. 1991. *Simians, Cyborgs, and Women: The Reinvention of Nature.* London: Free Association Books.

Ideland, Malin. 1997. Kroppssamhället: Om genetikens metaforer. *Kulturella Perspektiv* 1/97.

Ingold, Tim. (1988) 1994. *What is an Animal?* London and New York: Routledge.

Jay, Martin. 1993. *Downcast Eyes: The Denigration of Vision in Twentieth-Century French Thought.* Berkeley: University of California Press.

Kamratposten, 1995, no. 17.

Keller, Evelyn Fox. 1987. *Refiguring Life: Metaphors of Twentieth-Century Biology.* New York: Columbia University Press.

Lakoff, George, and Mark Johnson. 1980. *Metaphors We Live By.* Chicago: University of Chicago Press.

Lundin, Susanne. 1996. Longing for Social and Biological Identity: Parenthood by Means of Biotechnology. In *Force of Habit: Exploring Everyday Culture,* ed. Jonas Frykman and Orvar Löfgren. Lund: Lund University Press.

Lundin, Susanne. 1997. *Guldägget: Föräldraskapet i biomedicinens tid.* Lund: Historiska Media.

Lundin, Susanne, and Lynn Åkesson. 1996. Creating Life and Exploring Death. *Ethnologia Europaea* 26:1.

Martin, Emily. 1991. The Egg and Sperm: How Science has Constructed a Romance Based on Stereotypical Male-Female Roles. *Signs* 16.

Martin, Emily. 1994. *Flexible Bodies: Tracking Immunity in American Culture from the Days of Polio to the Age of AIDS.* Boston: Beacon Press.

Melucci, Alberto. 1996. *The Playing Self: Person and Meaning in the Planetary Society.* Cambridge: Cambridge University Press.

Merchant, Carolyn. 1989. *The Death of Nature: Women, Ecology and the Scientific Revolution.* San Francisco: Harper & Row.

Oudshoorn, Nelly. 1994. *Beyond the Natural Body: An Archaeology of Sex Hormones.* London and New York: Routledge.

Papagaroufali, Eleni. 1996. Xenotransplantation and Transgenesis: Im-moral Stories about Human–Animal Relations in the West. In *Nature and Society: Anthropological Perspectives,* ed. Philippe Descola and Gísli Pálsson. London and New York: Routledge.

Petchesky, Rosalind Pollack. 1987. Foetal Images: The Power of Visual Culture in the Politics of Reproduction. In *Reproductive Technologies: Gender, Motherhood and Medicine,* ed. Michelle Stanworth. Cambridge: Polity Press.

Rapp, Rayna 1993. Accounting for Amniocentesis. In *Knowledge, Power, and Practice: The Anthropology of Medicine and Everyday Life,* ed. Shirley L. Lindenbaum and Margaret Lock. Berkeley: University of California Press.

Ristilammi, Per-Markku. 1995. Optiska illusioner – fetischism mellan modernitet och primitivism. *Kulturella Perspektiv* 1995:3.

Rose, Hilary. 1994. *Love, Power and Knowledge: Towards a Feminist Transformation of the Sciences.* Cambridge: Polity Press.

Sachs, Lisbeth. 1996. *Sjukdom som oordning: Människan och samhället i gränslandet mellan hälsa och ohälsa.* Stockholm: Gedins.

Strathern, Marilyn. 1992. *Reproducing the Future: Essays on Anthropology, Kinship and the New Reproductive Technologies.* Manchester: Manchester University Press.

The Indipendent, 1 May 1996.

Turner, Bryan S. 1994 "Preface". *In The Consuming Body,* ed. Pasi, Falk. London: Sage.

Wheale, Peter, and Ruth McNally (eds.). 1995. *Animal Genetic Engineering: Of Pigs, Oncomice and Men.* London: Pluto Press.

Understanding Gene Technology through Narratives

Malin Ideland

Genetics and gene technology are sciences which affect 'ordinary people' in the highest degree. The experts do not have exclusive rights to knowledge of the human being; scientists and non-scientists appear to be in agreement about this. Yet this is not easily accessible knowledge. How many of us really understand that every little cell in our body can contain a string of DNA that is two metres long? And who can really grasp how the scientists on the HUGO project can succeed in locating the gene for prostate cancer or Huntington's chorea? How is it possible to see these?

In my ongoing doctoral research, I am studying how the mass media contribute to information about this medical science, how they make complex scientific arguments comprehensible and culturally manageable.[1] Since the mass media represent the major part of the information about genetics and genetic engineering, they also create an arena for the construction of an everyday ethic for these topics. This is where people acquire the information on which they base their opinions and feelings about the new biotechnology.

Different methods are used to explain and arouse emotions about modern medical technology. Many journalists, for example, frequently use metaphors to illustrate the laws of genetics. The body is described in societal terms, genes are given human properties, cells are described as factories, and the immune system as a defence (Ideland 1997; Martin 1994). In this way, the reader/viewer/listener can envisage the internal workings of our body. Another way to culturalize genetics is to describe body processes or genetic manipulation in pictures. Pictures are effective both for explaining a process and for playing on emotional strings. Horrifying pictures of genetically manipulated animals can be excellent weapons in the struggle against genetic engineering.

This article, however, will not primarily be about the news media, but rather about feature films. For a third way to handle a complex reality is to make fiction of it. Films and comic strips which frighten people with genetic engineering, organ donation, or test-tube fertilization arise as science makes increasing progress and new medical technologies become common in everyday consciousness, perhaps above all as a topic of debate.

I believe that popular culture is significant in the formation of people's opinions. Besides entertaining, feature films have at least two functions in culture: they are a mirror of society and a source of information and understanding about subjects like gene technology (cf. Nelkin and Lindee 1995; on the functions of different media, see also Hannerz 1990). This means that one must take into consideration the message that is spread through this channel as well. The films are American – often very American – but they are relevant for Swedish culture and everyday life when they enter people's living-rooms thanks to television and video.

Films can also be compared with legends and fairy tales from pre-industrial society (Ideland [Svantesson] 1995). In both present and past, folklore functions as a commentary on the contemporary cultural and social situation, contributing to a collective consciousness. It reflects dreams and threats, and it often serves as a safety valve for fears. In the world of film or story, serious matters such as genetic manipulation can be presented in extreme terms, making the consumers of fiction stop and think. This article deals with today's and yesterday's folklore, the cultural ideals that it reflects, so that we can see what has changed and what has remained the same.

Contextual Interpretation

The folktale of 'The Fisherman and his Wife' tells about the poor fisherman who caught a magical flounder which begged to be let go in exchange for granting the wishes of the fisherman and his wife. The wife begins by wishing for a new house, but she gradually becomes more and more avaricious. She wishes for a castle and gets one; she becomes queen and empress and is given increasingly lavish palaces. Finally, however, there is nothing more on earth to wish for, so the fisherman calls on the flounder once more and tells it that his wife wants to be God and have all the power under the sun in her hands. At this, all the things she has been granted go up in smoke – literally: the castle burns down and the flounder is never seen again.[2]

This story reflects a society in which God was considered to be the great authority, whose power no one could share. No human being was supposed to covet that power or try to challenge the almighty God. This theme recurs in many stories, including *Frankenstein* by Mary Shelley (1995). This classic horror story is about the young scientist Victor Frankenstein, who wanted to blow life into a man-made person, since he thought that death was unfair. He thus tried to advance science, but his research had disastrous consequences, for his monster lacked a soul and hence also the competence to feel empathy. Victor Frankenstein was later forced to pay a high price for trying to outwit death and acquire divine powers.

It was no coincidence that the tale of Frankenstein appeared in the cinemas once again in the winter of 1994–95. The director, Kenneth Branagh, said himself that the film should be viewed as a comment on developments in biotechnology (Söderqvist 1995). The idea that man should not assume too much power – although God has now been replaced by Nature – survives in (post)modern folklore. Stories about what happens when we manipulate Nature are infinite. They occur not just in the world of film and fiction, but also in other media contexts, such as articles in the tabloids. The folklorist Ulf Palmenfelt (1995) has compared this form of media expression – 'folklorized reality' – with the legends of bygone times. As in the world of fiction, the boundaries here are better defined than they are in reality. It is obvious who is good and who is bad, what one may and may not do. This is therefore a good source of inspiration when one is to shape one's own opinions, one's private everyday ethic. It is easier to take a stance when one knows what is right and what is wrong.

As culture has changed, the authorities in the narratives have been replaced, as we have seen. This transformation is reflected in Michel Foucault's (1973) statement that health and modern medicine have replaced salvation and the clergy in the modern era. The Western world view has changed, and religion has had to give way to science. This is also reflected in folklore.

In folktales such as 'The Fisherman and his Wife' or in 'Death as Godfather' (Liungman 1949) – in which a future doctor at his christening receives death as his godfather, and with it the gift to see instantly whether a patient is going to survive or not – God or the Devil is the authority, with the power to decide about life and death. In the modern folklore that I am studying, doctors, scientists, and genetic engineers have taken over the role of both God and the Devil. They have assumed power over life, and sometimes they are depicted as good people, sometimes as bad. Often both poles occur in the same film, a good and an evil scientist.

It may seem somewhat crude to equate doctors with God/Satan. God and the Devil acted as moral figures, but it is doubtful whether scientists can be viewed in this way. They should perhaps rather be compared with the itchy-fingered people in the stories who seek to challenge divinity, which in the past was represented by God and the Devil but nowadays by Nature. For it is Nature that is the supreme authority today. In the world of film, it often strikes back, punishing the scientists who have tried to manipulate it, and restoring the natural order. We see this, for instance, in the film *Jurassic Park*, where a millionaire enlists scientists to clone dinosaurs as an attraction in a theme park. The consequences are disastrous, since the prehistoric animals do not belong in a twentieth-century context. They are just one result of people trying to acquire too much power, so Nature rebels against the people who have tried to manipulate it; the Tyrannosaurus Rex and the other dinosaurs refuse to let themselves be tamed.

The ethnologist Inger Lövkrona (1996: 105–106) argues that the interpretation of a narrative is constituted in the social practice where discourse and experience meet. This means that folklore often has different meanings depending on the context in which it is narrated. A narrative about genetic manipulation is frightening because it reflects a realistic scenario, something that can actually happen. It is also frightening because it challenges one of the sacred things of our time, Nature (cf. the discussion by Bråkenhielm and Westerlund in this book). In another context, for example, pre-industrial peasant society, it was much more dangerous to challenge the religious authorities, and it is against this background that pre-modern folklore should be interpreted.

Folklore as a Comment on Society

Topical matters on the scientific agenda also become interesting in folklorist contexts. If we look at the cinema repertoire of the last few years, we can see clear links with new research. In recent years there have been films (and more are on the way) involving genetics and genetic engineering in one way or another. It is often a matter of cloning, presumably because this technique both frightens and amuses people in its absurdity. It can be horrifying, as in Ira Levin's book (later a film) *The Boys from Brazil* (1991), or when the dinosaurs in *Jurassic Park* run amok. On the other hand, it is amusing in the film *Multiplicity* (1996), where the character played by Michael Keaton clones himself in order to get more out of life. Cloning can also be used as a rational explanation for why characters come back to life in sequels, as Sigourney Weaver's character in *Alien* will soon do.

Gene technology has also been used in films about infectious and deadly diseases, with power-hungry scientists producing viruses using gene technology. And in *Species* (1995), American military research manages by genetic engineering to create a girl who is half-alien.

It is not only genetic engineering, however, that has succeeded in tickling the fancy of film producers. Similar stories have been invented about other techniques associated with medicine, human beings, and their bodies. From the end of the 1980s, for example, we find films such as *Robocop* (1987) and *Terminator* (1984 and 1991), the theme of which is humans with computerized bodies. Robots with human looks and human properties began to seem like a realistic possibility when research into virtual intelligence began. The Danish philosopher Peter Kemp (1991: 241–244) says that these robot-people challenge the idea of the irreplaceability of humanity. If it is possible to produce human-like beings, then man may become superfluous and easily replaced. The fear of a technique like cloning can be understood in the same way: it means that the individual is replaceable, or at least the idea exists.

With the progress of modern transplant surgery, the film world also began to take an interest in organ donation. The questions of where the soul resides, and whether personality can accompany the organ, were raised with films like *Heart Condition* (1989) and *Body Parts* (1988). In the latter, a man is given a new hand, transplanted from a convict on death row. The hand lives a life of its own, it is violent, and the formerly so mild-tempered and loving man starts hitting his children and almost strangles his wife with his new hand. The personality of the murderer thus comes with the organ.[3]

Not even this theme is new, however. As in the idea that one should not challenge God or Nature, there is here a tenacious structure in the view of transplanted organs. This is seen, for example, in the folktale of the three doctors who had organ transplants. In an attempt to prove their skill to each other, one of them removed a hand, one an eye, and one the stomach, and then they went to bed, the idea being that they would put the organs back in place the following day. As luck would have it, a dog ate up the organs during the night. The maid and manservant who had failed in their duty of guarding the organs found a way out. They got a new hand from a beheaded thief, an eye from a cat, and a stomach from a pig. When the unsuspecting doctors operated on themselves to put the organs back into place, they also acquired the characteristics of the former owners. The thief's hand wanted to steal, the cat's eye was always on the watch, and the pig's stomach wanted to eat everything that it came across, including dung.[4]

After transplants, the film makers began to take an interest in sperm donation, test-tube fertilization, and surrogate mothers. This theme is seen, for instance, in *Junior* (1994), and *Made in America* (1993). As in the films about transplantation techniques and ro-bot-men, there are warnings here against manipulating reproduc-tion processes.[5] Yet this is nothing new either. Old folk beliefs in-cluded a multitude of prohibitions and rules of behaviour which pregnant women had to follow. If they did not, there was a risk that the child would be born deformed or with the wrong character: it could be a cripple or a thief. If a pregnant woman took part in the butchering, for instance, the result could be that the child was born with the falling sickness – epilepsy. In the world of fairy tales, how-ever, the consequences could be even more dire. When a childless queen received help from an old sorceress in her eagerness to have a child, she was advised to eat two well-peeled red onions. She for-got to peel one of the onions, and as a result she gave birth first to a son in the shape of a serpent (*lindorm*) and then to a beautiful prince (Liungman 1949: 89ff.).

In today's Swedish culture it may seem that the prohibition on manipulating reproduction is not as strong as it used to be. The state permits abortions and contraceptives, and it subsidizes IVF treatment. Even in pre-industrial peasant society, however, there were also methods for influencing reproduction. Illegal abortions were performed, mothers did away with unwanted children, couples used coitus interruptus and various methods to try to influence the sex or the looks of the baby (Ideland 1995; Tillhagen 1983). The dif-ference is that none of this was permitted; the state and the church prohibited all manipulation of reproduction.

Swedish legislation in the field is still relatively restrictive; for example, egg donation is not permitted, as it is in many countries. And if we study modern folklore, we find there too a certain scep-ticism about the new technologies of reproduction. In the film *Made in America*, for instance, a warning is issued about accepting do-nated sperm without knowing where it came from, and without knowing anything about the genetic heritage that one is passing on to one's children. In *Junior* we see not only how a child is born by breaking the laws of nature, but also how gender boundaries are transgressed. With the aid of genetic engineering, a man manages to become pregnant and then gives birth to the child. This of course causes a great deal of problems, and we see how 'unnatural' it is (see further Dofs-Sundin 1995). In other films, such as *The Unborn* (1991), the question is raised of how far we can trust doctors and scientists, how sure we can be that a foetus has not been manipulated, that it really is the baby that is expected.[6]

Folklore in the form of film gives a constant commentary on contemporary society, thus also contributing to the perception of society.

Normalizing Narratives

The gallery of characters in folklore should likewise be interpreted in their cultural context. In the modern film folklore that I am discussing here, there is always a scientist in the cast. Geneticists, doctors, and molecular biologists are portrayed as being more or less crazy, totally obsessed with their research. In their longing for money and fame, or in a few cases in a desire to do something good for the world, they take their experiments too far. They exceed the limits of what is permitted, and the consequences are terrifying.

The picture of the mad scientist already existed in the nineteenth century, for instance, in the form of Frankenstein, Dr Jekyll, and other literary figures, who are still invoked in the criticism of to-day's medical experiments. The scientist is virtually always a man,[7] which may seem self-evident since the medical field is dominated by men. In the old fairy tales and legends, however, it is women, sometimes in the form of witches, who have power over reproduction and the body. With the professionalization of medicine, this power was shifted from the female sphere to become a male competence (see, for example, Merchant 1989: 172).

In Western culture, woman has been considered to be closer to nature than man, who is associated with culture (ibid.: 17ff.). Whereas the cultured man has had his place in the working life of society, reproduction has been the lot of the natural woman. This dichotomy also recurs in modern folklore. The men denaturalize culture through genetic engineering and other manipulations of the body, while women stand for normality and restoring things as they should be. In *Jurassic Park* it is the young woman who stops the rampage of the dinosaurs by outwitting the computer system, and in *Junior* a woman who gives birth in the natural way shows us the 'proper' sex roles (Dofs-Sundin 1995). Similarly, in the tale of the prince who is born as a serpent, it is in the end a girl who sees what he really is and helps him out of his bodily prison so that he can become a 'normal' person.

Folklore always presents a well-defined view of what is normal/natural/right or abnormal/unnatural/wrong. It has an educational function, showing what one may or may not do, for example, with the 'natural' body, or what the 'normal' sex roles are. Or it instructs people to respect the irreplaceability of man and the laws of nature, and not to clone oneself or dinosaurs. Folklore also helps to maintain cultural categories, such as gender, race, accepted forms of reproduction, or the difference between life and death. It does so by challenging these categories and then showing what is right. Foucault (1989) has shown how we justify and normalize a state by means of segregating practices. By defining a phenomenon as unnatural, one simultaneously assigns the quality of natural to the opposite. The narration of folklore of this kind thus has a segregating function.

Yet there is a paradox in this provocation of fixed categories; while they are maintained by means of the challenge, they also become frightening as a result of it. Here, I think, is the source of a great deal of the fear of genetic engineering, that it challenges these phenomena which we wish could be permanent and non-manipulable (see also Ideland 1996).

Playful or Serious?

Are the consumers of folklore unable to distinguish between fiction and reality? Have films and fairy tales any real significance in the information about genetics and genetic engineering? Is it not just playful escapism?

The ethnologist Lotten Gustafsson (1995: 6) argues that people at play apply a double consciousness. No matter how enthusiastically they indulge in play, people interpret the world around them from two perspectives, that of reality and that of play. She nevertheless believes that sometimes these consciousnesses blend, so that the boundary between reality and fiction is erased.

I think that the same thing happens with folklore. Its consumers can distinguish between fiction and reality. Just as little as people really believed that princes could be born as serpents, no one today believes that men can give birth or that adults can clone full-grown copies of themselves. Films and stories are interpreted in terms of the dual consciousness. Yet the rhetoric and the ideas are incorporated in the culture. In the more realistic mass media, such as newspaper articles and television programmes, parallels are often drawn to the world of fiction. Here too, *Jurassic Park* and *Multiplicity* serve as examples. It is to these that journalists, and no doubt many others, relate when cloning is discussed. In this way, the narratives are incorporated to become a part of a larger narrative about what genetics and gene technology really are. Folklore, whether true or not, thereby becomes part of the arena from which people seek inspiration in the construction of an everyday ethic for these issues.

This blend of the playful and the serious is nevertheless seen as a problem by many of those who possess a large symbolic capital and hence also the possibility to define what is to be considered knowledge and what can be called fiction. Lotten Gustafsson also describes how the arrangers of Medieval Week in Visby get upset when the participants mix frivolous elements in this serious re-enactment of medieval life. In the same way, I have heard doctors, scientists, and medical journalists complain about the inadequate information and the poor knowledge of genetics and gene technology among the general public. They are not considered to have the competence to distinguish between myth and reality. I think, however, that one must interpret the myths and their infiltration into the popular narrative of biotechnology as the critical comments of the people on what is going on in the sciences.

To sum up, it may be said that reality is incorporated in fiction and vice versa. Folklore tells stories about genetics and genetic engineering, based on contemporary society. At the same time, it comments on and criticizes modern medical technique, partly on the basis of cultural ideals with long traditions. It also has an educational task of helping to define normal and abnormal and in informing people about gene technology. Moreover, it infiltrates other narratives about genetics, thus helping to make the new biotechnology culturally manageable.

Notes

1 My dissertation has the working title *Genetic Dreams and Threats* and is part of a larger project, 'Genethnology: Genetics, Genetic Engineering, and Everyday Ethics'.
2 The Swedish version of this story is taken from Waldemar Liungman's collection *Sveriges samtliga folksagor i ord och bild* (1949: 229).
3 On the problems associated with the personality of the donor accompanying the donated organ, see, for example, Sachs 1996; Lundin and Åkesson 1996; Åkesson 1997.
4 The tale comes from Liungman (1949: 263) and has also been analysed in Lundin 1996a.
5 For further discussion of the myths of IVF treatment, see Lundin 1996b.
6 This fear recurs in interviews with couples who have undergone IVF treatment (Lundin 1996b).
7 In the film *Body Parts* it is a woman.

References

Åkesson, Lynn. 1997. *Mellan levande och döda: Föreställningar om kropp och ritual.* Stockholm: Natur och Kultur.

Dofs-Sundin, Monica. 1995. Den födande mannen: Moderskap eller fa(de)rlighet. *Häften för kritiska studier* 4/95.

Foucault, Michel. 1986. *The Birth of the Clinic.* London: Routledge.

Foucault, Michel. 1989. *Madness and Civilization: A History of Insanity in the Age of Reason.* London: Routledge.

Gustafsson, Lotten. 1995. Den förtrollade zonen: Leken som möjlighet och fara under medeltidsveckan i Visby. *Kulturella Perspektiv* 2/95.

Hannerz, Ulf. 1990. Genomsyrade av medier: Kulturer, samhällen och medvetande av idag. In *Medier och kulturer*, ed. Ulf Hannerz. Stockholm: Carlsson.

Ideland [Svantesson], Malin. 1995. Genteknik och vardagsetik: Arbetsrapport. Department of European Ethnology, Lund University. (mimeo)

Ideland, Malin. 1996. Mördarmöss och signalsubstanser. Om genteknikdebatten i massmedia. Seminar paper. Department of European Ethnology, Lund University. (mimeo)

Ideland, Malin. 1997. Kroppssamhället: Om genetikens metaforer. *Kulturella Perspektiv* 1/97.

Kemp, Peter. 1991. *Det oersättliga: En teknologietik.* Stockholm: Symposion.

Levin, Ira. 1991. *The Boys from Brazil.* New York: Bantam Books.

Liungman, Waldemar. 1949. *Sveriges samtliga folksagor i ord och bild.* Part 1. Stockholm: Lindfors bokförlag.

Lövkrona, Inger. 1996. Suktande pigor och finurliga drängar: Erotisk folklore och konstruktionen av kön i det förindustriella Sverige. In *Åtskilja och förena: Etnologisk forskning om betydelser av kön,* ed. Britta Lundgren, Inger Lövkrona, and Lena Martinsson. Stockholm: Carlsson.

Lundin, Susanne. 1996a. Människans biologiska reserver. In *Samtider, Svenska Dagbladet* 4 October 1996.

Lundin, Susanne. 1996b. Power over the Body. In *Bodytime: On the Interaction of Body, Identity, and Society,* ed. Susanne Lundin and Lynn Åkesson. Lund: Lund University Press.

Lundin, Susanne, and Lynn Åkesson. 1996. Creating Life and Exploring Death. *Ethnologia Europaea* 26:1.

Martin, Emily. 1994. *Flexible Bodies: Tracking Immunity in American Culture from the Days of Polio to the Age of AIDS.* Boston: Beacon Press.

Merchant, Carolyn. 1989. *The Death of Nature: Women, Ecology and the Scientific Revolution.* San Francisco: Harper & Row.

Nelkin, Dorothy, and M. Susan Lindee. 1995. *The DNA Mystique: The Gene as a Cultural Icon.* New York: W. H. Freeman and Company.

Palmenfelt, Ulf. 1995. Den folkloriserade verkligheten. In *Nostalgi og sensasjoner: Folkloristisk perspektiv på mediekulturen,* ed. Torunn Selberg. Turku: NIF.

Sachs, Lisbeth. 1996. *Sjukdom som oordning: Människan och*

samhället i gränslandet mellan hälsa och ohälsa. Stockholm: Gedins.

Shelley, Mary. 1995. *Frankenstein or the Modern Prometheus.* Cologne: Könemann Verlagsgesellschaft.

Söderqvist, Jan. 1995. Fritt fall för Frankenstein: Kenneth Branagh tar sig stora friheter med Mary Shelleys roman. *Svenska Dagbladet* 13 September 1995.

Values in Gene Technology from a Clinical Perspective

Maria Anvret

Clinical molecular genetics is entering a new era which means new ways of thinking and handling all the obtained information.

The HUGO project is generating new data concerning our genome, which is further used in order to understand how certain diseases and risk factors are transmitted. This knowledge can be used to predict whether an individual is going to develop a disease or not.

The technology is developing and more questions can be answered. Today there are nearly no limits – less material is used, in fact one single cell, and the desired can be given. In some instances unwanted information will be released.

The PCR (polymerase chain reaction) amplification method revolutionized this field by having genetic material from a single cell amplified into millions of copies and visualized on an agaros gel. This PCR method opened the possibilities to obtain information if requested within a couple of hours.

Today our knowledge about the human genome and what it stands for is much higher than five years ago. At that time inheritance and risk calculations within families, where a specific trait was inherited, were performed by drawing a pedigree and obtaining the family history.

Risk calculations could to a certain extent be given from the inheritance pattern. Cytogenetic analyses were used as a complement in some cases where chromosomal aberrations were thought to be the explanation for the diagnosis. The family history was of key value at that time in order to be able to predict or draw any conclusions at all.

The panorama is different today. If the disease is known within the family and the gene sequence or chromosomal localization is known, one may in most cases only need DNA from the proband (the person who asks for the information) to be able to perform a genetic analysis. If the disease causing mutation is identified, the test result will be 100% certain.

Knowing this can in some cases create problems for the proband. If the disease can be treated then there is something to offer but if it cannot and the genetic risk of 25% or 50% suddenly is changed to 100% or 0% it is more dramatic. From this it is obvious that genetic counselling is a prerequisite for genetic testing. The proband has to get all the available information and possibilities to ask all questions which might come up. Depending on the test and the position in the pedigree, we might also indirectly release information about other family members. The counsellor has to be aware of this and can often predict it.

More and more genes responsible for specific genetic disorders and genes predisposing for certain traits are characterized. New information is obtained on a daily basis about our genes. This allows new conditions to be diagnosed and predictions can be made in some cases.

The obvious question to arise is the following: 'How should we deal with all this information – can we change the future?' I don't think we can change anything. The techniques are here and will be further developed together with the competence.

Our responsibility is to handle all this information in the best way, in other words, not doing more than we are asked or doing what is relevant to do based on the clinical information about the family.

From this we will learn more and get better insight into the phenotypic variations among individuals with the same clinical diagnosis. By knowing this, a better understanding of how to treat the patients and possibilities to perform gene therapy in the future will be generated.

The wheel has started to rotate and the best we can do is to handle all this information in the best way and to be open in discussions.

Modern Biotechnology – Sustainability and Integrity

Law, Public Opinion and Politics in the Norwegian
Regulation of Modern Biotechnology

Torben Hviid Nielsen

The Ambiguities of Modern Biotechnology

'Modern biotechnology'[1] is a unique condensate and ambiguous mix
of knowledge and engineering, science and technology, nature and
culture, possibilities and risks, hopes and fears. Even less than the
other collections of artefacts, design and knowledge usually named
'technologies', it is far from being one single and unified technology. A set of uniform and interrelated techniques with an open field
of diverse applications appears a more adequate description or conceptualization.

Knowledge and Engineering

Modern biotechnology has already been institutionalized as closed
'black boxes' in a series of uniform and repetitive laboratory routines. Its procedures are looked up in manuals and can be used
without a command of the knowledge and principles involved. And
with only minor modifications, the same few sets of procedures are
used on micro-organisms, plants, animals and human beings; on

germ and body cells; to clone identical individuals or to fabricate transgene species, as well as for diagnostic purposes, in order to cure diseases, to 'improve' healthy individuals, etc.

The path from knowledge to engineering and from genetic blueprint to social behaviour turned out to be less straightforward than James Watson expected when he proclaimed that 'if you can study life from the level of DNA, you have a real explanation for its processes' (Watson 1995). Yet modern biotechnology has been and is still often understood and represented as a kind of 'biological information technology' (Keller 1995). DNA is 'code', 'text', or 'information'. DNA techniques are recoding. And the Human Genome Project (HUGO) promises to lift the veil of the 'code of codes' (Kevles and Hood 1992). Modern biotechnology may thus have changed the way we think about life more than the way we actually live our lives. In a new wave of biological determinism the genes are often presented as necessity, cause or fate (Hubbard and Wald 1993). Genes stand versus society, heritage versus environment, the necessity of DNA versus the freedom of the individual, and biological predestination versus social responsibility.

The discrepancy between the determinism of the scientific programme and the multiple and diverse applications of the techniques is unique to 'modern' biotechnology, and not fully grasped by the two dominant schools in contemporary studies of science and technology. 'Technological determinism' is transcended by the technology's multiple and diverse applications and 'social constructivism' by its dependence on nature.

Between Nature and Culture

The products of modern biotechnology are often conceptualized as 'hybrids', 'cyborgs' (Haraway 1985), 'monsters' (Law 1991) or 'quasi-objects' and as 'history' or 'culture' as much as 'nature'. No untouched or authentic nature is left behind. The 'biotechnologized' world accomplishes 'the end of the antithesis between nature and society', which began with the first industrial revolution (Beck 1992: 80ff.). 'Nature' and 'society', according to Bruno Latour (1993: 85ff.), no longer have any other existence than East and West: both are but relative points of reference in our attempt to classify the world. Biotechnology's intervention has added a stability dimension to the nature–society dimension. The ontology of quasi-objects is variable, and the apparent 'schism' between nature and society has been transformed into a delayed expression of stability.

Latour might overgeneralize the scenarios and overvalue their consequences. Yet modern biotechnology's recoding of nature challenges traditional 'Western' understanding of technology and nature. Identical procedures are applicable to micro-organisms, plants, animals and human beings – and to germ as well as body cells. Most procedures presuppose 'discovered' nature as well as 'invented' technology, and it is often impossible to attain one without the other and difficult to separate them even on a conceptual level, and the relationship between procedure and product is equivocal. The same products can be attained by different biotechnological procedures, and some products can be attained by way of traditional methods such as breeding as well as by modern biotechnology.

Techno-fix or Dystopia?

It has been argued that modern biotechnology is but the extension of the engineering profession's old dream of the 'techno-fix' from physics and chemistry to biology and thus life itself (Bud 1993: 12). 'For 3.5 billion years the genes took their own course with only minor interventions. Today it is a possible question whether it will be Homo sapiens – rather than the time-tested machinery of evolution – that writes the next chapter in the history of the genes?' (Suzuki and Knutson 1992: 2). The programme is, however, grandiose, ambiguous and it met with diverging attitudes. Modern biotechnology is presented as a threatened civilization's new hope. It will feed the world while restoring nature exhausted by traditional cultivation back to sustainability (cf. *Agenda* 21, ch. 16). And with the coming of gene therapy, cures for otherwise incurable and terminal diseases will be offered. The potentialities are, however, perceived by others as false hopes or even dystopia. The new knowledge might turn out to be an unopened Pandora's box, where only hope is left on the bottom. And the technology might create Frankenstein's monster, the unintended, unforeseen and uncontrollable evil.

Potentialities and risks, hopes and fears are probably more diverse than for any other 'modern' technology. The image of Janus, the two-headed Roman god, may depict modern biotechnology more adequately than any other technology.

Norwegian Biotechnology Legislation

Modern biotechnology was a controversial public and political issue from its very beginning. The often told story of the scientists' own moratorium in 1974 is a marked contrast to the secretiveness surrounding the military development of atomic technology at the end of the Second World War. The key issue in all regulation of modern biotechnology is to balance between utilizing possibilities and fostering hopes on the one side and eliminating or reducing risks and fears on the other. Part of biotechnology's techniques are regulated by the application of old, more general laws, but most industrialized nations have passed new laws regulating the most controversial procedures and applications.

Two acts passed by the Norwegian Parliament regulating the non-human and human use of biotechnology represent a unique and interesting national solution to the general problematic. The legislation followed a prolonged public debate and several White Papers, but was finally timed to pass Parliament before the second referendum on membership in the European Union on 28 November 1994. 'The Genetic Engineering Act' passed Parliament on 2 April 1993 and 'The Act Relating to the Application of Biotechnology in Medicine' followed on 14 June 1994.

Types of Regulation and Legal Forms

The two Norwegian acts both intend to balance between utilizing potentialities and avoiding risks and concerns by way of 'conditional approval'. They regulate on the level of the product, the process as well as the entire research programme, i.e. the three agendas described by Sheila Jasanoff (1995) as dominating American, British and German culture. And both phrase the conditions in three legal forms.

The 'legal standards' or general clauses are special to the Norwegian acts. The 'Genetic Engineering Act' states 'the aim to ensure that the manufacture and use of genetically modified organisms takes place (1) in an ethically and socially justifiable way, (2) in accordance with the principle of sustainable development and (3) without detrimental effects on health and the environment'.[2] And the 'Act Relating to the Application of Biotechnology in Medicine' states the need 'ensure that the application of biotechnology in medicine is utilized in the best interests of human beings in a society where everyone plays a role and is fully valued. This shall take place (1) in accordance with the principles of respect for human dignity, human rights and personal integrity, (2) without discrimination on the basis of genetic background and (3) based on ethical norms relating to our western cultural heritage'.[3] The two sets of legal standards are both 'standards of prevention' extending protection and responsibility from individuals to collectives and allowing for a substitute representation of interests. Their content and implications are, however, of a different nature. The act regulating non-human use has its terminology from recent socio-political and scientific discourse, whereas the act regulating use on humans borrowed its terminology from eighteenth-century natural law and human rights declarations.

A second legal form common to the two acts is the enumeration of detailed 'casuistic' requirements and preconditions for contained and exposed use of genetically modified organisms, artificial insemination, foetal diagnosis, genetic mapping, gene therapy, etc. Minor differences exist, especially in the classification of genetically modified organisms. Whereas the legal standards were specific to the Norwegian acts, the 'casuistic' requirements are basically identical with the requirements of the corresponding EU directives.

Both acts, finally, delegate a considerable influence over the interpretation and application of the legal standards to the 'Norwegian Biotechnology Advisory Board'. The board opens up for negotiations, formation of consensus or a balance of interests or compromises, and the circumstances and interests considered legitimate are indicated by the board's composition of professionals, public officials and interest representatives.

To sum up: The Norwegian legislation is thus characterized by special legal standards as well as the combination of these legal stand-

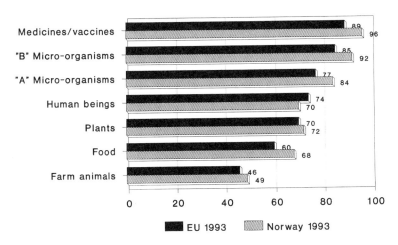

Figure I. Support for specified biotechnology research.

ards with the 'casuistic' preconditions essentially identical to the EU directives, and the influential advisory board as an administrative procedure. The legal forms do not equalize the domains of regulations, but together the three legal forms regulate biotechnology's products, procedures as well as the entire research programme.

Public opinion

A Eurobarometer survey from 1993 (Eurobarometer 39.1), which was also undertaken in Norway (Nygård 1995), indicates a basic harmony between the restrictive Norwegian legislation and a sceptical public opinion.

The Anthropocentric Hierarchy of Nature

The ambiguity and multiple applications of the technology are mirrored in the fact that public support for biotechnology research is heavily dependent on its intended purpose as well as on the organisms used.

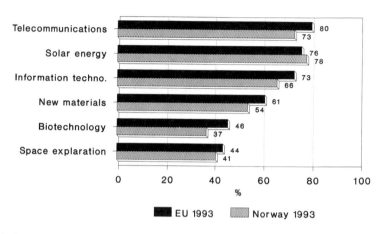

Figure II. Anticipated effects of new technologies.

Nine out of ten EU citizens find research for the development of new human medicine worthwhile, while less than four out of ten find research on domestic animals to improve agricultural output worthwhile. Highest support is generally given to research which is to benefit 'human beings' and is carried out on 'lower' organisms. Public opinion thus combines the two criteria in what adds up to be an 'anthropocentric hierarchy of nature' (cf. Figure I).

Low Expectations of Improvements in Daily Life, but High and Increasing Acceptance of Research and Development

In spite of the high support given to specified and well-defined research purposes, general expectations and anticipated effects of biotechnology on 'our way of life' are low.

EU citizens rate biotechnology second from the bottom of six new technologies, and Norwegians rate biotechnology at the bottom of the six, even lower than space research (cf. Figure II). Expectations of biotechnology divide the Norwegian population into three al-

Fig. III. Anticipated Effects of Biotechnology rank % 'will improve' – 'will make worse'	
Spain	55
Portugal	49
Italy	45
France	37
Great Britain	37
Ireland	36
Belgium	31
Luxembourg	31
Greece	25
Northern Ireland	25
Denmark	22
Eastern Germany	21
Netherlands	15
Western Germany	12
Norway	9

Figure III. Anticipated effects of biotechnology, national ranking.

most equally large thirds. Just over a third (36%) are 'optimists' about biotechnology, i.e. they expect that it will improve our way of life. A little more than a third (37%) either 'don't know' (28%)

General Acceptance of Biotechnology
Research. 1979, 1991 and 1993.
% In favour - against

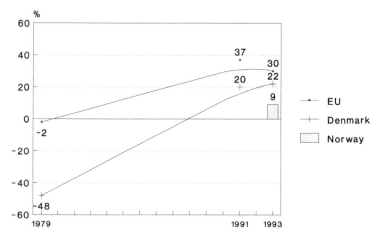

1979: Worthwhile - uacceptable risk
1991/3: Will improve - make things worse

Figure IV. General acceptance of biotechnology research, 1979, 1991 and 1993.

or expect 'no effect' (7%). And less than a third (27%) are 'pessimists', i.e. they expect biotechnology to make things worse.

Norway fits nicely into the general picture, showing expectations
of biotechnology to be highest in the mostly Southern European
nations, where it is also least prevalent, and lowest in the Central
and Northern European nations, where research and use are also
most prevalent.

Spaniards express the highest expectations: 61% expect biotechnology to improve their way of life, only 6% to make their way of
life worse, giving a balance of opinion of +55. Norway, with 36%
'optimists' and 27% 'pessimists', has the lowest balance of opinion,
i.e. +9. The national ranking contradicts the stereotype of a restrictive, Catholic South and a liberal, Protestant North. At least part
of this national ranking might alternatively be explained by an 'in-
tuitive' public understanding of the 'declining marginal utility'.

Expectations of biotechnology's influence on everyday life are low,

Fig V. 'Cognitive' Knowledge of Biotechnology rank after % correct answers	
Denmark	62
Netherlands	59
Great Britain	58
France	56
Norway	54
Luxembourg	53
Western Germany	51
Belgium	50
Eastern Germany	50
Italy	46
Northern Ireland	45
Ireland	45
Spain	44
Greece	39
Portugal	39

Figure V. 'Cognitive' knowledge and anticipated effects of biotechnology research, national ranking.

relative to other new technologies as well as absolutely. Public acceptance and expectations of biotechnology research have nevertheless increased rapidly during the two decades, as the technology has moved from the laboratories over the moratorium to the marketplace (cf. Figure IV).

As the increase has been the lowest in nations where support al-

"Cognitive" Knowledge and Anticipated
Effects of Biotechnology.
Norway, 1993

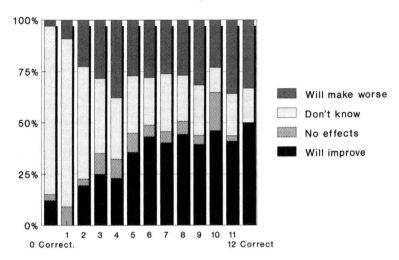

Figure VI. 'Cognitive' knowledge and anticipated effect of biotechnology research, Norway, 1993.

ready was strongest, a tendency to homogenization of the nations at the same higher level is distinct.

Knowledge, Expectations and Acceptance

Scientists, industry, and research politicians often consider lack of knowledge as a major cause of low expectations and acceptance. 'If people knew more, they would also expect and accept more.'

Expectations and knowledge do however rate the European nations in a nearly reverse order (Figures III and V).[4] High knowledge and low expectations co-exist in the same nations – and vice versa.

A positive, although low, correlation between knowledge and expectations is to be found, however, within most nations.

Norwegian expectations thus rise quite steeply and at the same rate as cognitive knowledge up to the level slightly above the average knowledge (cf. Figure VI). Above this level the increase in expectation is met by a corresponding decline in the number who 'do

Fig. VII		
"The Proponent" and "the Opponents" Profiles		
Marked over-representation (Cf. Appendix II)		
The "Proponents"	The "Opponents"	
The Entrepreneur	The "Blue" Critique	The "Green" Skepticism
M		
"Best" age	Elder	Younger
Higher education	Primary school	Higher education
Larger city	Rural, village	Larger city
Conservative, Labour Party	Political centre	Political left
Non-religious	Religious	Non-religious
"Materialist"	"Materialist"	"Post-materialist"
Large knowledge	Little knowledge	Large knowledge

Figure VII. Profiles of the 'proponents' and the 'opponents', Norway, 1993.

not expect any effect' or 'do not know'. And among the most knowing half of the population the number of 'pessimists' increases nearly as much as the number of 'optimists'.

The Entrepreneur, the 'Blue' Critique and the 'Green' Scepticism

The demographic, socio-economic and value characteristics of the 'optimists' and 'pessimists' illuminate the background for the political alliance behind the legislation as well as the continued struggle over the innovation and spread of biotechnology. The one out of three Norwegian 'optimists' is a well-defined group, sharing the characteristics that general innovation theory expects to find among the technological entrepreneurs (cf. Figure VII and Appendix II).

The typical 'optimist' is a young man with long education and high income living in a major city; he votes Conservative followed by the Progress Party, and Labour; his values are 'non-religious' and 'materialistic'; his knowledge of biotechnology is good, he considers his own influence strong, and he perceives few risks in biotechnology.

However, a cluster analysis of the one out of three Norwegian 'pessimists' shows that they are made up of two distinct and equally large segments of the population. The first segment has a marked over-representation of elderly, with a lower level of education, living in a rural environment, voting for the political centre, considering themselves 'religious' and 'materialistic' and with a relatively small knowledge of biotechnology. The typical representative of the second segment is a younger female, with long education living in a major conurbation. She shares socio-economic background and good knowledge with the optimists and proponents, but votes for the political left and considers herself 'non-religious' and 'post-materialistic'.

The two segments of opponents represent respectively a 'blue' or 'value-conservative' critique and a 'green' or 'modern' scepticism, and each amounts to one out of six Norwegians. The 'blue' voices the Christian or Faustian critique: even if biotechnology succeeds as technology, it is a covenant with Mephistopheles. The 'green' voice the modern fear of new risks as described in Ulrich Beck's Risk Society.

The Politics of Norwegian Biotechnology Law

The two Norwegian biotechnology acts followed an extensive public debate and a prolonged legislative process. Parliament itself set the wheels rolling in May of 1987, when it asked for 'a White Paper concerning the ethical guidelines for research and development of biotechnology and genetic engineering'. The concern of the parliamentary majority was that biotechnology could transgress 'natural' or 'social' boundaries in an undesired and/or unintended way. The majority asked for ethical guidelines because they considered ethics as basically a right to say 'no', a potential 'veto', which – *vis-à-vis* the risky technology – could legitimately suspend routine politics.

From Politics via Ethics Back to Politics
The prolonged legislative process ended up as a circular movement and a kind of 'trial and error process' from the politically formulat-

ed concern via attempts at ethical delineation back to the quoted legal standards of a more political nature.

The first major official policy document discussed in the horizon of 'normative' philosophic ethics. It contrasted consequences with emotive arguments, 'utility' with 'duty', and it referred to classical philosophers such as Jeremy Bentham (1748–1832) and Immanuel Kant (1724–1804).[5] The second official policy document focused instead on the 'descriptive' values and attitudes of the population as indicated by opinion polls. 'Ethical guidelines' were interpreted as a 'set of moral principles', on which there should be 'broad agreement … among the general public whatever their political or religious beliefs'.[6] In the third attempt, however, the interregnum Conservative[7] and the Labour government[8] both altered the terrain referring to the new concept of a 'mixed ethics', which was said to solve the conflict between 'utility' and 'duty', and to have the greatest acceptance among the people – i.e. to be normatively as well as descriptively true.

The three successive attempts all left traces in the preliminaries to the final acts, but none are crucial as key concepts in the final texts. Finally the politicians replaced the attempts to define 'ethical' conditions, which they had asked for, with legal standards of a more political nature.[9]

The legal standards designate or anticipate from a mere formal point of view new technological opportunities, which cannot be regulated by legal casuistics. The preliminaries leave no doubt, however, that they were also intended as substantial and restrictive extra requirements compared to the subsequent casuistry as well as the EU directives. However, as the differences between the 'casuistics' in the Norwegian acts and the EU directives are only minor, a more 'liberal' or 'permissive' interpretation and administration cannot be ruled out. The three legal forms thus stretch across a latent conflict. Parliament intended the specific Norwegian legal standards to be substantial extra requirements, but if administered as such the EU or EEA may claim that they in reality are but a concealed barrier to trade.

Beyond Left and Right

In spite of its punctual timing, the Norwegian government did not succeed in keeping the biotechnology legislation unaffected by the EU agenda. The two major political issues were marked by the same departure from traditional demarcations between 'left' and 'right', 'change' and 'conservatism', 'modern' and 'traditional', etc. (cf. Giddens 1994) The neo-liberal Progress Party (FrP), the Conservative (H) and Labour (Ap) were the EU-proponents and most in favour of biotechnology. The value-conservative Christian People's Party (KrF), the Centre Party (S) and the Socialist Left Party (SV) were the EU opponents and most sceptical of biotechnology.[10]

Considering this new parliamentary alliance pattern, the survey indicates a surprisingly good harmony between party politics and the electorate's attitudes. The profile of the proponents' electorate reflects the parties who voted for the biotechnology acts, as the profiles of the 'blue' critique and the 'green' scepticism reflect the parties that voted against the acts. The Labour Party and the Conservative Party jointly formed the nucleus of the supporters in terms of party politics as well as electorates. The opposition includes the Centre Party as well as the opposite poles on the traditional left–right axis, the value-conservative 'blue' critique voiced by the Christian People's Party and the modern 'green' scepticism voiced by the Socialist Left.

Notes

1   For an overview see Fransman and Roobeck 1995.
2   'The Act Relating to the Production and Use of Genetically Modified Organisms (Gene Technology Act)'. Act. No. 38 of 2 April 1993. Ministry of Environment. Official translation. Cf. also Backer 1995.
3   'The Act Relating to the Application of Biotechnology in Medicine'. Ministry of Heath and Social Affairs. Official translation. Oslo, Norway. June 1994.
4   'Cognitive knowledge' is an additive index of 12 questions (cf. Appendix I). The validity of this index and a previous attempt in the Eurobarometer 35.1 is discussed in the INRA report to the European Commission on the Eurobarometer 39.1.
5   Norges Offentlige Utredninger 1990:1. 'Modern Biotechnology. Health, Safety and the Environment'. See especially VII.
6   Norges Offentlige Utredninger 1991:6. 'Man and Biotechnology'. See especially IV. Quotations from p. 44.

7   Department of Environment, Parliamentary Proposition no. 8 (1990–1991), 'Concerning Biotechnology'.
8   Department of Environment, Parliamentary Proposition no. 36 (1990–91), 'Additional Proposition Concerning Biotechnology'.
9   For a Swedish discussion see Forsman and Welin 1995.
10 During the final vote in Parliament six amendments were proposed to the 'Genetic Engineering Act', four of them jointly from the Christian People's Party, the Centre Party and the Socialist Left Party. During the final vote for the 'Act Relating to Medical Use' twenty-six amendments were proposed. The Christian Peoples Party supported twenty, the Centre Party seventeen and the Socialist Left twelve, whereas the Conservatives supported four and the Progress Party three.

References

Agenda 21: Programme of Action for Sustainable Development. Rio Declaration on Environment and Development. United Nations, 1992

Backer, Inge Lorange. 1995. 'Sustainability' and 'Benefits to the Community' Concerning the Release and Use of Genetically Modified Organisms in the Norwegian Gene Technology Act. *Proceedings of the International Conference on Release and Use of Genetically Modified Organisms: Sustainable Development and Legal Control,* pp. 41–47. Published by the Norwegian Biotechnology Advisory Board. Oslo.

Beck, Ulrich. 1992. *Risk Society: Towards a New Modernity.* London: Sage Publications.

Bud, Robert. 1993. *The Uses of Life: A History of Biotechnology.* Cambridge: Cambridge University Press.

Eurobarometer 39.1. *Biotechnology and Genetic Engineering: What Europeans Think about it in 1993.* Report written for the European Commission by INRA. October 1993.

Forsman, Birgitta, and Stellan Welin. 1995. *The Treatment of Ethics in Swedish Government Commission on Gene Technology.* Studies in Research Ethics 6. Gothenburg: The Royal Society of Arts and Sciences in Gothenburg.

Fransman, Martin, and Annemieke Roobeck. 1995. *The Biotechnology Revolution?* Oxford: Blackwell.

Giddens, Anthony. 1994. *Beyond Left and Right: The Future of Radical Politics.* Cambridge: Polity Press.

Haraway, Donna. 1985. A Manifesto for Cyborgs: Science, Technology and Socialist Feminism in the 1980's. *Socialist Review* 1985: 65–107.

Hubbard, Ruth, and Elijah Wald. 1993. *Exploring the Gene Myth.* Boston: Beacon Press.

Jasanoff, Sheila. 1995. Product, Process or Programme: Three Cultures and the Regulation of Biotechnology. In *Resistance to New Technology*, ed. M. Bauer. Cambridge: Cambridge University Press.

Keller, Evelyn Fox. 1995. *Refiguring Life: Metaphors of Twentieth-Century Biology.* New York: Columbia University Press.

Kevles, Daniel J., and Leroy Hood. 1992. *The Code of Codes: Scientific and Social Issues in the Human Genome Project.* Cambridge, Massachusetts: Harvard University Press.

Latour, Bruno. 1993. *We Have Never Been Modern.* New York: Harvester Wheatsheaf.

Law, John (ed.). 1991. *A Sociology of Monsters: Essays on Power, Technology and Domination.* London and New York: Routledge.

Nygård, Berit. 1995. *Ny bioteknologi i Europa.* Rapport 1/1995. Trondheim: Senter for Bygdeforskning.

Suzuki, David, and Peter Knutson. 1992. *Genetics: The Ethics of Engineering Life.* London: Unwin.

Watson, James D. 1992. A Personal View of the Project. In *The Code of Codes: Scientific and Social Issues in the Human Genome Project*, ed. D. J. Kevles and L. Hood. Cambridge, Massachusetts: Harvard University Press.

Appendix I

'Cognitive' Knowledge

Norway 1993

	Right	Wrong	Do not know
There are bacteria which live from waste water (**True**)	85%	2%	13%
It is possible to find out whether a child will have Down's Syndrome (i.e. will be a 'mongol') within the first few months of pregnancy (**True**)	83%	5%	12%
Biotechnology/genetic engineering makes it possible to increase the milk production of cows (**True**)	77%	5%	19%
Yeast for brewing beer consists of living organisms (**True**)	75%	7%	18%
It is possible to change the hereditary characteristics of plants, enabling them to develop their own defence against certain insects (**True**)	64%	8%	29%
Children look like their parents because they have the same red blood cells (**False**)	60%	13%	27%
It is possible to modify bacteria genetically so that they will produce useful substances (**True**)	49%	10%	41%
The cloning of living things produces exactly identical offspring (**True**)	50%	10%	40%
There are test tube babies who were developed entirely outside the mother's body (**False**)	39%	19%	42%
Genes of all living things on earth are made up of different combinations of only 4 or 5 chemical blocks (**True**)	19%	13%	69%
Most bacteria are harmful to human beings (**False**)	43%	43%	14%
Viruses can be contaminated by bacteria (**False**)	19%	36%	45%
	54%	14%	31%

Appendix II

SUPPORT FOR RESEARCH AND DEVELOPMENT
Analyze of Variance (ANOVA)
NORWAY, 1993

	ETA	BETA		ETA	BETA
Sex and Age			**'Values'**		
Men	.68	.67	Conservative (H)	1.52	1.48
Women	−.70	−.69	Neo–liberal (Frp)	.97	.93
Sex	12	12	Labour (Ap)	0.40	.37
18–25 years	−.13	−.23	Socialist Left (SV)	−.73	−.57
26–40 years	.58	.61	Centre party (S)	−2.76	−2.74
41–55 years	00	.01	Christian Party (KrF)	−3.37	−3.46
56–79 years	−.68	−.64	**Political Party**	27	27
Age	08	08	Non-religious	.35	.34
'Social–economic background'			Religious	−.45	−.44
Working class	−.02	−.01	**Religion**	07	07
Middle class	−.23	−.20	Materialist	.44	.21
Upper class	1.83	1.49	'Mixed'	.13	.13
Subjective Class	08	08	Post–materialist	−.19	−.92
Primary school	−.18	.13	**'New' Values**	09	07
Secondary school	−.03	−.12	Little Knowledge	−.53	−.71
University	.45	−.12	Large Knowledge	.38	.51
Education	04	02	**Knowledge**	08	11
Oslo, Bergen, Trondheim	.80	.73	Small Risk	1.17	1.12
City	.24	.23	Great Risk	−1.22	−1.17
Village and Countryside	−1.46	−1.38	**Risk perception**	21	20
Urban–rural	14	13			
R = .208			**R = .285**		

Biotechnology and the European Public

Björn Fjæstad

Biotechnology is sometimes referred to as 'the next nuclear power issue', in reference to a widespread and consistent public resistance. There are several reasons for this. One is that biotechnology comprises a number of very powerful techniques and methods in quite different areas of application (e.g., human and veterinary medicine, pharmaceuticals, agriculture and food processing, environmental cleansing, chemical industry); another reason is that the debate about risks has not yet fully engaged the wider public or the popular media; a third is that a successful introduction of these processes and products most likely will need the approval of the public – as voters, taxpayers, consumers, members of interest groups, and employees – especially regarding perceived risks in relation to perceived benefits, an acceptance that is by no means a matter of course. Recent research suggests that public opinion has progressively gained importance at earlier stages of the introduction of new technology.

The background is the increased public awareness of the effects of technological implementation in society. Until not too many years ago, the players involved were mainly scientists, industry, banks and investors, governments and government agencies, and international institutions. Decisions about new technological systems were made by those elite groups in society whom, ideally, the public, the shareholders, or some other constituency had entrusted with making decisions. Very little public discussion took place about the pros and cons of different technologies. However, mainly since the debates on nuclear power in most Western countries in the 1970s and 1980s, public resistance and public acceptance are increasingly part of socio-technological change.

The study discussed here aims to describe and analyse the evolution of the national debates about and the public perception of modern biotechnology, particularly genetic engineering.

The Project

The research project Biotechnology and the European Public involves about 15 research groups from as many countries,[1] and is made up of several interrelated parts. Its general goal is to highlight the social aspects of the introduction and reception of genetic engineering into our various countries, with a wide meaning given to the word social: including legal, political, psychological, and economic aspects. The time period covered is from 1973 through 1996.

The project started officially on 1 January 1996. It is coordinated from London by a group of researchers at the Science Museum, The Imperial College, and The London School of Economics, with Professor John Durant as head coordinator.

Each national team carries out studies in three different arenas of public discourse. They are

- the policy arena
- the media arena
- the arena of public opinion.

Each team has an obligation to provide the international comparative study with a minimum set of results and analyses: in some cases a number of time series, and in other cases a description of, for instance, the debate that took place prior to the first legislation concerning release of genetically modified organisms.

All the national studies are nationally financed, except for the empirical data collection of the 1996 Biotechnology Eurobarometer, which is paid for by the EU. The EU has also provided a concerted action grant, making possible a few international workshops each year.

The national research groups are staffed in a variety of ways, with varying numbers of people, from different academic fields, and with varying degrees of seniority. All in all, more than thirty people are actively involved and meet at the project's international workshops.

In a multinational, multi-study project like this, of course, a number of choices have been made. The complex web of 24 years of events, processes, and situations has had to be reduced to something that can actually be compared across countries, and this is one of the reasons for the decision to work with the three relatively well-defined arenas. The following short descriptions of the arenas refer mainly but not exclusively to the Swedish study.

The Policy Analysis

The goal of the policy module is to map and understand the national debates on genetic engineering over the years, the initiatives made, the actors and their objectives and, of course, the final outcomes in the form of actual policy, for instance legislation or voluntary agreements among various players.

It is an extensive task to make an inventory of all the relevant government committee reports, the political party platforms, the lobbying actions of the industry, and so on. The investigators are looking for the major events and initiatives.

In the Swedish project we are now busy collecting printed material from various sources, and we are also in the process of interviewing scientists, administrators, members of parliament, lawyers, and others who were central and instrumental to the chain of developments during our period of study.

To make it all manageable, we initially focus upon a relatively small number of questions:

- Who were the actors and what were their motives?
- What were the issues and how were they framed?
- What were the policy outcomes and why?
- Was there any public participation in the policy process?

The policy arena is the most difficult one for the comparative analysis, since the national end results are a set of narrations and not a number of quantitative measurements. However, a graphic method will be developed for the national policy debates to be condensed into time diagrams showing activities of various kinds and intensities.

The Arena of Public Opinion

In all the 15 EU member states and in Norway, a Eurobarometer survey of Biotechnology was fielded in November 1996. The Canadian survey was carried out in February 1997. The questionnaire was the result of work in our project, even if the EU had the final say, since it was the EU that actually commissioned the study from the polling institutes. The questionnaire was quite different from the previous Eurobarometer, for instance with more emphasis on applications. We have just started to analyse the data, and the results are embargoed until a press conference in Brussels later this year.[2]

However, some preliminary attitudinal work was carried out in several countries in preparation for the Eurobarometer survey. In Sweden, we conducted focus group interviewing and we administered a mail survey to a sample of 800 representatively chosen people. One conclusion from these studies is that there really is no such attitude object as 'genetic engineering'. People obviously have quite different attitudes to different applications, and they are not even certain of which applications really are genetic engineering and which are not.

In Sweden, the public opinion arena is handled by the doctoral candidate Susanna Olsson. In the exploratory work, she found, on the whole, that our respondents tend to be negative to or at least quite cautious of applications of genetic technology. They also seem to have a very moderate amount of factual knowledge.

The preliminary studies further indicated that the public tends to become more negative, the closer applications get to humans, with the exception of some really urgent ways of using the technology, such as treating severe hereditary diseases.

The attitudes do not seem to be based on knowledge, but mainly to be generalized from other areas and from deeply felt emotions regarding, for instance, the holiness of life or the risk of creating a society where babies are made to order. Interestingly, physical risks seem to make much less difference than the moral implications of a particular application of genetic engineering. As could be expected, women, younger people, and persons to the left of the political centre tend to be more negative than men, elder persons, and people to the right of centre. Politically, those with the most negative attitudes are, not surprisingly, those who support the Green Party, and those with the least knowledge are the Christian Democrats.

Besides this mainly cross-sectional survey of perceptions carried out through the Eurobarometer, similar studies carried out since 1973, including single questions asked in other studies, are located and discussed in the project.

The Media Analysis

In this module, some concrete choices have been made, and they are not exactly identical in all countries. But there is a substantial core that is quite comparable. The reasons for the relative diversity are two: the media structure varies from country to country, and the access to archives is also dissimilar.

The basic requirement for each country is to analyse one paper medium, i.e., the one newspaper or news-magazine that is the most important agenda-setter in the country. For instance in the US it is the *New York Times*, in Britain *The Times* up to the mid-1980s and then *The Independent*, in Germany *Der Spiegel* and so on. In Sweden the choice is easy: *Dagens Nyheter* is the largest and the leading morning newspaper in the country.

Sweden has a rather typically mixed situation regarding the access to archives. Until 1992, *Dagens Nyheter* manually cut clippings from the actual newspaper and filed them in one or a few of several thousand different categories. Since 1992, a digital on-line archive has been in use.

We have examined all manual clippings in about 80 different categories and have included in our population all stories dealing with 'the intervention, handling, and/or analysis at the level of the gene(s)'.

No limit is set on size or type of editorial material. For instance, we have included articles that obviously refer to genetic engineering without explicitly mentioning it. Some of the editorial leaders, as well as some cultural, economic, and debate articles, are of this nature.

In some countries, especially those of Latin and southern Europe, it is maintained that the media coverage of genetic technology really cannot be separated from the coverage of IVF and other reproductive techniques. And thus, some of our colleagues have included all reproductive technology in their media samples. In Sweden, we have chosen not to do so, since in our understanding there is not the same confluence between the two debates here. But, of course, we have included all stories on reproductive technology or reproductive ethics that fall within our definition.

All in all, we have more than 600 articles in our population. The analysis is still in its early stages, but it is possible to report here, for the first time, some preliminary results. There are four periods when the number of articles peaks: 1979, 1983, 1989, and 1995. Interestingly, each peak is consecutively higher. The first one dealt mainly with safety for laboratory personnel and for people living close to the research facilities; there was a perceived risk of leakage. There were also calls for another moratorium and for restrictive legislation.

The 1983 peak seems mainly to have concerned ethics in general and the diagnostic use of genetic engineering in particular. In 1989, the focus was on regulation and, also, on ethics. The 1995 peak has not been analysed yet.

Considering only stories with a transgenic direction, there is an interesting development. The 1979 peak dealt almost exclusively with micro-organisms, in 1983 there were about 50 per cent more stories on micro-organisms than on humans, and in 1989–90, the stories on human transgenics had become three times as numerous as stories on any other type of organism.

Throughout the studied period, in-house journalists account for a majority of the stories. But there is a tangible number of other authors as well. Interestingly, in 1979 and 1983, scientists were the most common non-journalist group to write articles. In 1989, the topic of genetic engineering had obviously become established since the most common group of authors, after in-house writers, were wire agencies with politicians as runners-up.

During all the study period, better health is by far the most common benefit mentioned. Other benefits include economic progress, a cleaner environment and specific uses for the Third World.

On the other hand, there has been a shift in the newspaper coverage of risk. In 1979 the emphasis was on safety for laboratory workers and for the public, but in both 1983 and 1989 the most publicized risk by far concerned moral and ethical issues.

From 1973 through 1992, out of more than 400 stories, only 44 contained an explicit positive valuation while 82 expressed an explicit negative valuation. The negative articles were considerably more negative than the positive ones were positive. Generally, the 1983 articles seem to have been more positive towards genetic engineering than those from 1989.

Anna Olofsson, another doctoral student, is in charge of the Swedish media study. She has also carried out a special, exploratory analysis of the 63 articles from 1989, one of the top years. The general picture is that of negative or neutral stories, with the debate focused on legislation and regulation, or rather the lack of it. Ethics was seldom the main theme, but was frequently a second or third theme. A substantial part was straight reporting of scientific and technical developments. Largely, the articles were written by the newspaper staff. The actors focused on were mainly scientists but also the government and the political parties. The general public is totally absent both as actors and as contributors to the newspaper.

Only a small proportion of the articles are judged as being positive towards genetic engineering. Of the 63 articles, only 11 were positive, and 2 of them had genetic engineering as a side theme; 27 were negative. The negative articles were bigger, and quite a few of them were placed on the debate and editorial comments pages. None of the positive stories were placed there. Instead, they were all published under domestic news.

The negative stories were heterogeneous as regards writer, subject, and placement. The positive ones were, on the other hand, homogeneous. Some of the negative articles were written by external writers, in many cases by well-known personalities. Only one of the positive ones had an external writer, and this was a story from a news agency.

Finally, we are also doing a cross-sectional media study. In all the participating countries, we monitored a large number of media during the month of November 1996 for items on genetic engineering, with special reference to the soy and maize issues. The aim here is mainly to find out how different media frame genetic engineering, that is, what aspects are emphasized, what language and metaphors are used, and which consequences are highlighted.

The premise of our research project, Biotechnology and the European Public, is that public issues of technology evolve within an information environment. The two dominant systems involved are the policy discourse and the media discourse. They interact with each other and have consequent impacts on public opinion which can, in turn, exert important influences on media and policy activities. Our aim is to study and understand this triad with special reference to patterns of technology development and social change.

Note

1 Austria, Canada, Denmark, Finland, France, Germany, Great Britain, Greece, Italy, The Netherlands, Norway, Poland, Sweden, Switzerland, USA. Surveys are also carried out in Japan and New Zealand.
2 Since this talk, an overview of the results has been published in *Nature*, vol. 387, 26 June 1997, pp. 845–847.

Reference

Olofsson, Anna, and Susanna Olsson. 1996. The New Biotechnology: Media Coverage and Public Opinion. In *Public Perceptions of Science, Biotechnology, and a New University*, ed. B. Fjæstad. Mid Sweden University (Mitthögskolan), Report 1996:10.

International Comparisons of Values
Experiences from the European Values Studies
Loek Halman

Introduction

Values are regarded as important determining attributes of human behaviour. They are at the core of interpretations of modern advanced industrial society and the transformation processes which are taking place within societies (see e.g. Van Deth 1995: 2), and as such they are an important subject for (empirical) research. Within Europe the diverging or even contrasting values in the various countries can easily hamper the further unification of Western Europe. Value similarity is considered essential for the process of integration and the question what cross-national value differences exist in contemporary Europe is therefore highly relevant. The European Values Study (EVS) is an attempt to explore empirically the varieties and similarities in fundamental social, political, cultural, moral, and religious values held by the populations of the European countries. Numerous books, articles, and papers have appeared on the data from this survey project, gathered for the first time in 1981, and for the second time in 1990, delineating the value patterns of the European populations. The findings suggest that Europeans are far from homogeneous in their value patterns; there are significant differences in value orientations between the European countries (see e.g. Ester, Halman, and de Moor 1994).

It is not the intention of this paper, however, to dig deeper into the problems concerning the understanding of the varieties and similarities in patterns of value orientations. The purpose of the paper is twofold: (1) to discuss the concept of values and our approach to empirically assess people's value orientations; (2) to review the problems of comparative research with a special focus on the difficulties related to comparing value orientations cross-nationally.

The Concept of Value

The concept of value appears to be very difficult and complicated. There exists hardly agreement upon the description or definition of values. As is the case with other theoretical concepts, it is unclear what values are, how many there are, and how important they are in human life.

According to the EVS initiators, values function as prime guidelines in people's lives, and therefore they are worth investigating. However, even in EVS the concept of values is hardly ever discussed thoroughly. The difficulties of properly defining and describing the concept of values are clearly reflected in the sociological and psychological literature about values, revealing a real terminological jungle (Brandsma 1977). The definitions of values varies with the different disciplines. In moral philosophy, for example, or aesthetics, the concept of value is used in a rather normative sense. Values are regarded as the ultimate criteria to decide what is good and bad, beautiful and ugly, right and wrong. In daily usage, values also have such a normative connotation which appears very obvious in sayings like: a decline in values, a values crisis, the re-establishment of old traditional values, and so on.

In the many sociological and psychological efforts to define values, references are made to the study by Lautmann (1971) who mentioned no less than 180 different value definitions. To a large extent this conceptual confusion is grounded in the nature of values. One obvious problem in (social) research is that values can only be postulated or inferred because values as such are not visible or measurable in a direct way. As a consequence values are more or less open concepts.

Values are theoretical concepts and as such hardly based on an empirically grounded theory, which in turn yields a speculative theoretical approach (Brandsma 1977: 62). This also becomes very clear in the many theoretical efforts to distinguish the concept of values from closely related concepts such as attitudes, beliefs, opinions, and so on, which in their turn, are also of a theoretical nature (e.g. Kluckhohn 1959; Friedrichs 1968; Williams 1968; Rokeach 1973; Scholl-Schaaf 1975). However, all value theories seem to share the notion that values are somehow more basic or existential than all these related concepts. Attitudes, it has been argued, refer to a more restricted complex of objects and behaviours than values do (Reich and Adcock 1976: 20), whereas norms 'are specific prescriptions and proscriptions of standardized practices' (Cohen 1978: 77). This way of theoretical arguing assumes a more or less hierarchical structure in which values are more basic than attitudes, norms, etc. 'A value is seen to be a disposition of a person just like an attitude, but more basic than an attitude, often underlying it' (Rokeach 1968: 124). Holding a specific value means a disposition, a propensity to act in a certain way, but 'not a specific form of conduct' (Cohen 1978: 77). For instance, one can argue that the number of children in a completed family, using birth control pills, visiting a family planning clinic, signing a petition for (legalized) abortion, can be explained by one's attitude towards family planning (Ajzen and Fishbein 1980: 88). In turn this attitude and possibly other attitudes may be explained by a more basic value with respect to family, marriage, and sexuality. So, in theory, we can distinguish two different steps in explaining concrete human behaviour. First, we can postulate different attitudes, opinions, norms, etc., to explain several behavioural acts. Secondly, we can take the argument one step further in stating that all these different attitudes may theoretically be explained by a more general underlying guiding principle with a much wider scope. These more general guiding principles can be called 'values'.

Since values, attitudes, beliefs, norms, and opinions, are all theoretical concepts and not observable in a direct way, it is theoreti-

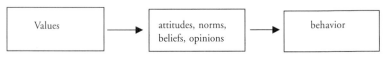

Figure 1. Values and related concepts.

cally as well as empirically difficult to distinguish between them. However, there is more or less consensus upon the idea that values can be seen as more basic, more deeply rooted, and less specific than opinions, attitudes, and so on. Values appear, so to speak, in these attitudes, opinions, and norms. This idea is indicated in Figure 1.

Our definition of values is a functional one and it is more a description of what values do rather than what they actually are. Values are seen as deeply rooted motivations or orientations guiding or explaining certain attitudes, norms, opinions which in their turn direct human action or at least part of it. Holding a specific value means a disposition, a propensity to act in a certain way (Halman 1991: 27).

Measuring Values

Values can be investigated in various ways. They can be observed by analysing documents or by observing concrete human behaviour. However, since the aim of the EVS study was to uncover the variety in values, beliefs, and attitudes of a great number of populations, and to discern whether these values and beliefs constitute patterns or systems, the only possible approach is the survey method. The EVS questionnaire was developed to tap values in important domains of life such as religion, morality, socio-economic life, politics, work, family, marriage, and sexuality. The selection of items in these domains of life was not so much guided by very well developed sociological theories in these domains such as the sociology of religion, political science, the sociology of work, and so on – which could have led to testing specific hypotheses within the theoretical framework of these disciplines. Instead, the selection of items was guided mainly by broad and very general ideas of theories of modernization.

The answers to the questions in the interview are assumed to reflect (at least partly) a person's value orientation. Consequently, knowing the answers to the questions, the underlying values can be discovered by exploring the common features in these answers. Such an approach implies of course that the content of the theoretical construct or value is assumed to be sufficiently determined by the items in the questionnaire.

In each of the domains mentioned, there was an attempt to trace the common feature in these behavioural responses – the answers to the questions in this domain – and to call this common feature a value. In empirical terms this approach demands a search for not directly observable, underlying, or latent variables.

Numerous statistical techniques are available and widely used to trace such underlying factors, such as factor analysis, latent class and latent trait analysis, Guttman scaling, cluster analysis, multidimensional scaling, and so on. These techniques have in common the assumption of the existence of a latent variable, and further the assumption 'that all the associations among the manifest variables can be explained by the dependency of these manifest variables upon the latent variable(s). In other words, when the latent variable is held constant, the manifest variables should be statistically independent. This basic assumption of the latent structure model as developed by Lazarsfeld and Henry is known as the assumption of local independence' (Heinen 1993: 6).

Since there was no theory guiding the development of the first questionnaire, there was no a priori idea of what to expect from the data. In EVS, factor analyses have mainly been used to explore the data, rather than to confirm certain a priori hypotheses on the data. By subjecting the European data to various factor analyses (principal components analysis), value patterns were discovered in the religious-moral domain, the social-political domain, the domain of primary relations (marriage, family, sexuality, and education), and in the domain of work (Halman, Heunks, De Moor, and Zanders 1987; Ester and Halman 1990; Halman 1991; Halman and Vloet 1994; Halman and De Moor 1994).

However, the results of such exploratory comprehensive analyses may be affected by a number of factors, such as varieties in the item formulation and format of the questions, and diversity in answer categories (ranging from important, yes–no dichotomies, Likert-like scales, to 10-point scales and choices between two statements), and so on. Therefore, the data have been re-analysed focusing on each dimension – each factor – which appeared in the exploratory analyses, separately. This facilitated the analyses, because the extraction of one factor only does not create difficulties about the number of factors to be extracted or the way of rotating the factor solution to an interpretable one (e.g. orthogonal or oblique). Further, techniques were used which are most appropriate to analyse the kind of data. In the case of dichotomous and non-interval data, factor analysis is not to be preferred (McDonald 1985; Lucke and Schuessler 1987). Instead latent class analysis or latent trait analysis are more appropriate.

The main differences between latent class analysis and latent trait analysis is the nature of the latent variable. In the case of latent class analysis, the latent underlying, not directly observable variable, is assumed to be a discrete, categorical variable with no order restrictions imposed upon its categories, whereas in the case of latent trait analysis, the latent variable is supposed to be, as in factor analysis, a continuum. In latent class analysis there are no further requirements concerning the relationships between manifest and latent variables, while in latent trait analysis it is assumed that the probability of giving a certain answer increases or decreases monotonously with the position one has on the latent variable (e.g. Langeheine and Rost 1988: 2; Shockey 1988: 20; Rost 1988: 150; Heinen 1993: 28–29). Despite differences in assumptions and requirements of the various latent structure techniques, a common feature is the construction of a latent variable.

One of the (additional) purposes of these kinds of analyses is the calculation of scores for each individual which on these latent variables indicating an individual's value orientation (Shockey 1988: 289; Hagenaars 1993; Hagenaars and Halman 1989; Lucke and Schuessler 1987). By calculating mean scores for the entire population of a country, the relative positions of the countries may be compared. The mean scores are thus indicative of the value orientations of the entire population of a certain country (Halman and Vloet 1994; Halman and De Moor 1994).

Comparing Values Cross-Nationally

An essential prerequisite of comparing the scores between countries is, of course, that these scores are indeed comparable. The question, however, is whether this is the case or not. Performing factor analysis on the combined data of all European countries does provide an overall patterning of values, but does not guarantee that such a pattern is valid in all countries. The same problems appear when comparisons in time are made. This is one of the most important and all too often neglected issues in comparative research: the problem of comparability.

This problem of comparability is, however, not only apparent in the stage of data analyses, it arises in all stages of a study as well: in the stage of conceptualization, in that of operationalization, and in the data-collecting phase

The problems of conceptualization have been concisely stated by Frey: 'Cross-cultural research demands cross-cultural concepts' (Frey 1970: 187). Phenomena vary from specific to universal as to their occurrence in the various countries. But even if their occurrence is universal, they may need a country- or culture-specific interpretation (Armer 1973). For example, the concept of 'being a liberal' has a different meaning in the United States from that in Western Europe (Inglehart 1977; Klingemann 1979). 'Democracy' has a different meaning in a totalitarian society than in a society with a long-standing democratic history. 'Bureaucracy' is again another example in this respect, for it cannot be investigated in non-differentiated societies. Even an apparently unproblematic concept as 'age' appears to be not comparable *per se,* and the same applies to items as 'degree of urbanization', 'income', etc. (Verba 1969; 1971). It is necessary to define very accurately the phenomena one wishes to study and then decide whether or not these phenomena can be considered as universally appearing in the countries involved in the survey (Warwick and Osherson 1973; Grimshaw 1973; Malpass and Poortinga 1986; Armer 1973). Because EVS is confined to countries belonging to the sphere of Western culture, and as the design of the questionnaire was a co-operative endeavour of researchers who had intimate knowledge of the societies involved, one might assume this first demand to be sufficiently met.

The operationalizations in the various countries need to be carefully considered. The indicators of a concept should be comparable between countries, which sometimes demands different indicators. Comparable, however, does not necessarily mean identical.

Warwick and Osherson have summarized the difficulties associated with the translation of items in six 'dimensions of equality' (Warwick and Osherson 1973: 28–30):

1 Lexical meaning: a word is lacking to denote a phenomenon
2 Grammatical meaning: the direct translation of a word can result in a different interpretation and meaning
3 Context: the order of items or answer categories can be illogical in a certain society
4 Response style: answer categories may influence the response
5 Salience: a word can be very important in one society, and unimportant in another.
6 Scale points: the scale points can be interpreted differently in different cultures, nations.

From this overview it will be clear that there can never be a foolproof guarantee that the same thing has been measured in all the countries (Almond and Verba 1970: 351). Despite an apparently accurate translation of the original questionnaire, problems still may occur concerning the meaning and interpretation of separate items. Further, in some countries the tendency to choose extreme answer categories may be stronger than in other countries. Differences in answers may be caused by what has been called 'response style' or 'differential loquacity'. A response style is 'a tendency to choose a response category, such as "yes" or "agree", regardless of an item's content' (Warwick and Osherson 1973: 25). Differential loquacity implies that one population is more eager to respond (Americans for instance) whereas other populations (Chinese people for example) are rather silent. Asian people are found to respond with what is called a 'courtesy-bias or hospitality-bias': these people want to please the interviewer and thus they agree with him regardless of the content of the question (Deutscher 1973: 173). In such cases the equivalence of the concepts is not questionable, but the answers or scores do not indicate a similar position on a variable.

To what extent the quality of the data has been influenced by these and other aspects of operationalization and translation is hard to say. The translation of the questionnaire was the responsibility of the researchers of each country involved in the project. As there were no indications of differences of response styles, the agreement was that there would be literal translation of the answer categories from the standard English questionnaire.

In the stage of data collection an important issue is whether or not a survey yields reliable data. In the Western European countries and North America the interview is a frequently used instrument and there are no clear reasons to expect respondents in these countries to give unreliable answers, although differences in the non-response rates to some questions create the impression that in some countries these questions are answered with a certain reluctance. It is of course possible that this reluctance to answer questions is culturally determined. It may be a cultural habit to have no opinion, as it may be a cultural habit to have an answer to any question, as seems to be the case for many Americans. People in America even have an opinion about a non-existing television programme (Brislin, Lonner, and Thorndike 1973: 61). This phenomenon is known as the 'I can answer any question bias' (Brislin 1986: 163). In the former socialist countries of Central and Eastern Europe it is not quite unlikely that, having lived all of their lifetime under a strongly repressive regime, some of the respondents will not give reliable answers to some of the questions (Armer 1973: 67).

The collection of the data has to be done in a comparable fashion in all countries, but sometimes this is not possible. In some countries one simply cannot draw representative samples, because official lists of individuals or households are not available. In those countries where suitable lists are available, respondents can be selected at random, while in those countries where such lists are not available, the selection of respondents can be based on quotas set by, for example, age, sex, and occupation on the basis of census data.

Finally, problems of comparability appear in the stage of data analysis. These problems have partly been mentioned already as problems arising in preceding stages: different response styles, reluctance to answer questions, translation errors, etc. As far as EVS is concerned, the problems in this stage of analysis have to do with the comparability of value orientations. In analysing the data one has to be sure that the values or latent variables are comparable across nations. One might argue that one can never be sure that one is dealing with comparable latent variables, because the questions asked in an interview are only a small collection of all possible questions that could refer to the latent variable that is found. And how do we know that if more questions were asked the conditions of comparability would still have been met? The argument is right, but can be raised against all statements of a theoretical nature in all sciences. The only conclusion to be drawn from this argument is that scientific knowledge is of a hypothetical nature and open to revision (see Halman and De Moor 1994: 29).

How does one decide whether or not the latent variables are equal or at least comparable in the various countries? It is not sufficient to conclude that in all countries the same structure appears when performing factor analysis in the countries separately, or to put it into other words, that the same manifest variables (or items in the questionnaire) are related to a latent variable in all countries. The same variables need to refer in a similar way to the same factor(s) in all countries. Only if this condition is met can it be concluded that the factor(s) and the scores on these latent variables are comparable. Put more technically, the relationships between the manifest and latent variables have to be the same in all the countries investigated. If this indeed is the case we can, following Przeworski and Teune (1970), speak of identical values and therefore conclude that the scores on the latent variable are, in fact, comparable.

The same statistical tools as can been used to tap values can be applied to test the comparability of the latent variables. Investigating the comparability of the value orientations which have been discovered by means of factor analysis, can be achieved by using the LISREL program (Jöreskog and Sörbom 1981). Two models are of interest now. The first tests the hypothesis that the number of factors is equal in all countries. If this hypothesis cannot be rejected the conclusion is that the values or latent variables resemble dimensions. In all countries the same manifest variables are referring to a latent variable which, however, needs a nation-specific interpretation because the relations between manifest and latent variables are still 'country-dependent'.

Another model deals with the factor patterns. It tests whether these patterns are equal across groups, in our case the countries. If such a model should fit the data, the relations between the manifest and latent variables are equal in all countries, and it may be concluded that the latent variables or values are comparable between countries. Similar questions concerning the comparability of latent variables have to be answered when other techniques are applied discovering values, e.g. latent class analysis and latent trait analysis.

The acceptance or rejection of a hypothesis depends on the value of the overall chi-squared goodness-of-fit measure. However, this measure is strongly affected by the sample size, with the consequence that the models will show a lack of fit (Bentler and Chou 1987: 97; Heinen 1993: 49). So, the information of the chi-squared test statistic should be evaluated very cautiously. A recommended statistic, which can be easily calculated and which is useful in the decision as to which model has to be preferred, is BIC; the Baysian information criterion (see Raftery 1986). Akaike has proposed yet another statistic for evaluating various models: Akaike's information criterion (AIC), and a modification of AIC was proposed by Bozdogan. As will be known, 'the correction for degrees of freedom can be seen as a penalty for choosing models with too many parameters ... [and] in order to obtain a more stringent penalty for overfitting and to get an index which is asymptotically consistent' (Heinen 1993: 52) the consistent Akaike information criterion (CAIC) was proposed.

Our analyses often demonstrated that indeed the same (factor-) structure may be found in each country, but that models with more restrictions assuming equal patterns (identical loadings or identical conditional probabilities) in all countries have to be rejected (Halman and Vloet 1994). This implies that the relationships between the manifest and the latent variables are not the same in all countries, and thus the underlying, latent variables cannot be interpreted in the same way. Hence, the conclusion is that the values need a 'country-specific' interpretation.

Because the EVS surveys have been carried out at two points in time, 1981 and 1990, the question has to be answered whether the values measured in 1981 are comparable with the values measured in 1990. Using the same statistical techniques (LISREL and so on) these questions can be answered, and above all comparable scores can be calculated.

Concluding Remarks

Values are conceived as deeply rooted motivations or orientations guiding human behaviour. Values are theoretical concepts, which makes it hard to draw clear dividing lines between values and other theoretical concepts such as attitudes, opinions, norms, and so on. Defining values in terms of underlying factors guiding human behaviour implies that, in empirical research, values can be detected through exploring the basic features underlying a wide variety of behavioural and attitudinal items. In analytical terms this implies that values can be detected by analysing the answers of respondents and try to discover the common feature(s) in these answers. In EVS questions concerning specific domains were included, such as religion and morality, socio-economics, politics, work, family, marriage and sexuality. Various techniques, known as latent structure models, are available to discover such underlying features.

A major topic of concern in international comparative research settings is the issue of comparability, and this issue is not limited to the stage of data analyses but occurs at all stages of a research project. However, basically, all (sociological) research is comparable, even research that is confined to one nation or one group. In fact 'no social phenomenon can be isolated and studied without comparing it to other social phenomena' (Øyen 1990: 4). It can be argued, therefore, that all 'sociology is comparative – that, as a matter of fact, sociology cannot be done without making comparisons' (Grimshaw 1973: 3). In a study which is confined to one nation only, a researcher has to face similar problems to those which a cross-national researcher is confronted with. 'The cross-national survey faces

all the problems of the national survey-problems of conceptualization, sampling, interview design, interviewer training and so forth.
There is, however, one major difference. In the cross-national survey all these problems ... are multiplied by the number of nations
studied' (Almond and Verba 1970: 349). If, as in EVS, there are
not only several countries involved but also multiple measurement-
points in time, the number of problems has to be multiplied not
only by the number of countries, but also by the number of meas-
urement-points in time!

References

Ajzen, I., and M. Fishbein. 1980. *Understanding Attitudes and
Predicting Social Behavior.* Englewood-Cliffs, N.J.: Prentice-
Hall.

Almond, G., and S. Verba. 1970. Some Methodological
Problems in Cross-National Research. In *Comparative
Perspectives*, ed. A. Etzioni and F. DuBow, pp. 349–364.
Boston: Little, Brown and Company.

Armer, M. 1973. Methodological Problems and Possibilities in
Comparative Research. In *Comparative Social Research:
Methodological Problems and Strategies*, ed. M. Armer and
A. D. Grimshaw, pp. 49–81. New York: John Wiley.

Deutscher, I. 1973. Asking Questions Cross-Culturally: Some
Problems of Linguistic Comparability. In *Comparative
Perspectives*, ed. A. Etzioni and F. DuBow, pp. 163–185.
Englewood Cliffs: Prentice-Hall.

Bentler, P. M., and Chih-Ping Chou. 1987. Practical Issues in
Structural Modelling. In *Common Problems/Proper Solutions*,
ed. J. S. Long, pp. 161–192. Beverly Hills: Sage.

Brandsma, P. 1977. *Het waardenpatroon van de Nederlandse
bevolking: Een verkenning naar de plaats van de categorie
"waarde" in de sociologische theorie en het onderzoek.* Dissertation, Groningen.

Brislin, R. W. 1986. The Wording and Translation of Research
Instruments. In *Field Methods in Cross-Cultural Research*, ed.
W. J. Lonner and J. W. Berry, pp. 137–164. Beverly Hills: Sage.

Brislin, R. W., W. J. Lonner, and R. M. Thorndike. 1973. *Cross-Cultural Research Methods*. New York: John Wiley.

Cohen, P. C. 1978. *Modern Social Theory*. London: Heinemann.

Ester, P., and L. Halman. 1990. Basic Values in Western Europe: An Empirical Exploration. Paper presented at the XIIth World Congress of Sociology, Madrid, July 9–13, 1990.

Ester, P., L. Halman, and R. de Moor (eds.). 1994. *The Individualizing Society*. Tilburg: Tilburg University Press.

Frey, F. W. 1970. Cross-Cultural Survey Research in Political Science. In *The Methodology of Comparative Research*, ed. R. T. Holt and J. E. Turner, pp. 173–294. New York: The Free Press.

Friedrichs, J. 1968. *Werte und soziales Handeln: Ein Beitrag zur soziologischen Theorie*. Tübingen: J. C. B. Mohr (Paul Siebeck).

Grimshaw, A. D. 1973. Comparative Sociology: In What Ways Different from Other Sociologies? In *Comparative Social Research: Methodological Problems and Strategies*, ed. M. Armer and A. D. Grimshaw, pp. 3–48. New York: John Wiley.

Hagenaars, J. A. P. 1993. *Loglinear Models with Latent Variables*. Newbury Park: Sage.

Hagenaars, J. A., and L. C. Halman. 1989. Searching for Ideal Types: The Potentialities of Latent Class Analysis. *European Sociological Review* 5: 81–96.

Halman, L. 1991. *Waarden in de Westerse Wereld*. Tilburg: Tilburg University Press.

Halman, L., F. Heunks, R. de Moor, and H. Zanders. 1987. *Traditie, Secularisatie en Individualisering*. Tilburg: Tilburg University Press.

Halman, L., and R. de Moor (1994). Comparative Research on Values. In *The Individualizing Society*, ed. P. Ester, L. Halman, and R. de Moor, pp. 21–36. Tilburg: Tilburg University Press.

Halman, L., and A. Vloet. 1994. *Measuring and Comparing Values in 16 Countries of the Western World*. WORC Report. Tilburg: WORC.

Heinen, T. 1993. *Discrete Latent Variable Models*. Tilburg: Tilburg University Press.

Inglehart, R. 1977. *The Silent Revolution: Changing Values and Political Styles among Western Publics.* Princeton: Princeton University Press.

Jöreskog, K. G., and D. Sörbom. 1981. *LISREL V: Analysis of Linear Structural Relationships by Maximum Likelihood and Least Squares Methods.* Uppsala: University of Uppsala.

Klingemann, H. D. (1979). Measuring Ideological Conceptualizations. In *Political Action: Mass Participation in Five Western Democracies,* ed. S. Barnes, M. Kaase et al., pp. 215–254. Beverly Hills: Sage.

Kluckhohn, C. 1959. Values and Value-Orientations in the Theory of Action: An Exploration in Definition and Classification. In *Toward a General Theory of Action,* ed. T. Parsons and E. A. Shils, pp. 388–433. Cambridge: Harvard University Press.

Langeheine, R., and J. Rost. 1988. Introduction and Overview. In *Latent Trait and Latent Class Models,* ed. R. Langeheine and J. Rost, pp. 1–7. New York: Plenum Press.

Lautmann, R. 1971. *Wert und Norm: Begriffsanalysen für die Soziologie.* Opladen: Westdeutsche Verlag.

Lucke, J., and K. Schuessler. 1987. Scaling Social Life Feelings by Factor Analysis of Binary Variables. *Social Indicators Research* 19: 403–428.

Malpass, R. S., and Y. H. Poortinga. 1986. Strategies for Design and Analysis. In *Field Methods in Cross-Cultural Research,* ed. W. J. Lonner and J. W. Berry, pp. 47–83. Beverly Hills: Sage.

McDonald, R. 1985. *Factor Analysis and Related Methods.* Hillsdale: Lawrence Erlbaum Associates.

Øyen, E. (ed.). 1990. *Comparative Methodology: Theory and Practice in International Social Research.* London: Sage.

Przeworski, A., and H. Teune. 1970. *The Logic of Comparative Social Inquiry.* New York: John Wiley and Sons.

Raftery, A. 1986. Choosing Models for Cross-Classifications. *American Sociological Review* 51: 145–146.

Reich, B., and C. Adcock. 1976. *Values, Attitudes and Behaviour Change.* London: Methuen.

Rokeach, M. 1968. *Beliefs, Attitudes and Values*. San Francisco: Jossey-Bass Inc. Publishers.

Rokeach, M. 1973. *The Nature of Human Values*. New York: Free Press.

Rost, J. 1988. Test Theory with Qualitative and Quantitative Latent Variables. In *Latent Trait and Latent Class Models*, ed. R. Langeheine and J. Rost, pp. 147–171. New York: Plenum Press.

Scholl-Schaaf, M. 1975. *Werthaltung und Wertsystem: Ein Plädoyer für die Verwendung des Wertkonzepts in der Sozialpsychologie*. Bonn: Bouvier Verlag Herbert Grundmann.

Shockey, J. W. 1988. Latent-Class Analysis: An Introduction to Discrete Data Models with Unobserved Variables. In *Common Problems/Proper Solutions*, ed. J. S. Long, pp. 288–315. Newbury Drive: Sage.

Van Deth, J. W. 1995. Introduction: The Impact of Values. In *The Impact of Values*, ed. J. W. van Deth and E. Scarbrough, pp. 1–18. Oxford: Oxford University Press.

Verba, S. 1969. The Uses of Survey Research in the Study of Comparative Politics: Issues and Strategies. In *Comparative Survey Analysis*, ed. S. Rokkan, S. Verba, J. Viet, and E. Almasey, pp. 56–106. The Hague: Mouton.

Verba, S. 1971. Cross-National Survey Research: The Problem of Credibility. In *Comparative Methods in Sociology: Essays on Trends and Applications*, ed. I. Vallier, pp. 309–356. Berkeley: University of California Press.

Warwick, D. P., and S. Osherson. 1973. Comparative Analysis in the Social Sciences. In *Comparative Research Methods*, ed. D. P. Warwick and S. Osherson, pp. 3–41. Englewood Cliffs: Prentice-Hall.

Williams, R. M. Jr. 1968. The Concept of Values. In *International Encyclopedia for the Social Sciences*, ed. D. L. Sills, pp. 283–287. New York: Macmillan.

Explaining Attitudes towards Biotechnology
Cognitive, Moral and Emotive Factors
Wim Heijs and Cees Midden

Introduction

After the development of techniques to modify DNA in the 1970s, the seemingly limitless possibilities to which these techniques could give rise, led several researchers to put forward proposals for safety measures. At the same time a public debate arose about issues of safety, ethics and policy (Durant 1992). Since technological developments influence the lives of millions of people, government policy should involve making early predictions about the effects of introducing new technologies, to enable adequate and socially responsible decision-making. Nuclear energy is a case in point: public uncertainty and concern can hamper a proper discussion and a well-balanced introduction (De Loor, Midden and Hisschemöller 1991). With regard to biotechnology also, economic and social benefits will only be realized if consumers are confident that products are safe, effective and ethically sound (Cantley 1987; de Flines 1987; Hoban and Kendall 1992). In the interests of an effective policy and market-introduction, it is vital that existing attitudes to biotechnological applications and their structure and backgrounds are studied.

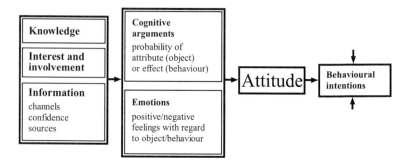

Figure 1. Schematic outline of the theoretical model.

Previous Research

Previous research showed that the public knows very little about biotechnology. The general view is a moderately positive one, but there is also some feeling of unease regarding the possible risks, especially concerning applications in agriculture and stock-breeding. Medical applications are regarded more favourably but moral objections exist where these applications are not used to save life. As a result of the lack of awareness, however, attitudes are diffuse and likely to be unstable. How attitudes would alter if the public was better informed is unknown. The same applies to any resultant behaviour, such as political views or the response to various biotechnological applications when introduced (Daamen et al. 1986; OTA 1987; Knulst and Van Beek 1988; De Loor, Midden and Hisschemöller 1991; Marlier 1992).

Past research had some weaknesses. First, the term 'biotechnology' appears too abstract. Because the public is not familiar with the term, questions eliciting a general response have produced answers with only a limited degree of validity. Secondly, the level of public knowledge on the subject has generally been tested by measurements of passive or associative recognition rather than a genuine underlying knowledge. Thirdly, there is insufficient insight into the underlying reasons behind judgements or the possible effects on subsequent behaviour.

Model

Figure 1 gives a schematic outline of the research model. The central concept is 'attitude'. Attitudes are presumed, more or less stable structures in the memory, representing acquired predispositions to react in a consistently favourable or unfavourable way to a specific object (concrete objects, places, people, behaviour or ideas) (Fishnein and Ajzen 1975).

Attitude theory is one of the most researched areas in social psychology. Attitudes have been modelled in various ways. According to the more recent views on the subject to date, attitudes are based on both cognitive arguments and emotive factors (Breckler and Wiggins 1989; Tesser and Shaffer 1990). Cognitive arguments are assessments of the probability that objects will have specific properties or that behaviour will have specific consequences (e.g. 'orange peel contains pesticide residue' or 'eating meat promotes health'). Emotions are positive or negative feelings evoked by the object. The relative contribution of cognitive arguments and emotions to the formation of attitudes is not necessarily evenly balanced. Some attitudes may have a more cognitive foundation, while others are more influenced by emotions. It is important to distinguish between the two, particularly in the interests of a public information policy.

Attitudes in turn evoke behavioural intentions, which can be interpreted as indications of future behaviour. In addition to attitude, behavioural intentions are also influenced by the opinions of reference groups, the desire to conform to these opinions and the estimation of one's own ability to carry out the behaviour (perceived control) (Ajzen 1991). Behavioural intentions are not perfect indicators of actual behaviour. For example, behaviour can be hindered by external factors or be prevented by a lack of ability on the part of the individual. Moreover, actual behaviour is usually influenced by a combination of different attitudes and behavioural intentions.

Cognitive arguments and emotions are not merely influenced by the object itself. They are also determined by individual psychological and structural characteristics. Important variables include background knowledge, interest in and involvement with the object. Knowledge may enhance the individual capacity to elaborate new information and argument. Personal involvement may motivate to a more critical consideration of arguments, which is the basis for a relatively stable and cognitively based attitude. In attempts to change these attitudes, e.g. by information campaigns, the content of the message is most important. Resistance to contradictory information is rather high. Lack of knowledge or interest generally means that an attitude is based less on the content and more on the context of the information, such as characteristics of the source or channel (e.g. credibility) (Petty and Cacioppo 1986; Breckler and Wiggins 1991). Less stable and more emotionally based attitudes may be the results.

Structural demographic characteristics can also influence the uptake of information in existing mental structures. For example, a higher educational level can lead to more cognitively based attitudes. It is assumed that these characteristics have only indirect effects on attitudes, which are mediated by the components of the model (Meertens and von Grumbkow 1988).

Research Questions

The research questions are: (1) What are public attitudes to biotechnological applications and to which cognitive arguments and emotions are these attitudes linked? (2) What are the levels of background knowledge and public interest in biotechnology? (3) What are the information sources, which sources enjoy public confidence and what are the gaps in public information? (4) What are the links between personal characteristics and cognitive arguments, emotions and attitudes? (5) What are the links between attitudes and behavioural intentions? (6) Can target groups be identified on the basis of these variables? (7) Are there trend-related changes?

Selection of Applications

Instead of using general indicators for attitudes towards biotechnology, a structure was developed to typify the large diversity of applications in a coherent way and to analyse the main evaluative dimensions. Previous research indicated that the following six aspects might influence the formation of attitudes to applications: (1) life-span (long, recent); (2) techniques used (traditional or new); (3) target objects (the organism the technique is directed at); (4) origin (of genetic material); (5) sector (agriculture, medicine, etc.); (6) end product/purpose. The public, of course, will not consider all of these aspects in the formation of attitudes. However, lay people do put forward arguments which suggest a consideration of more than one aspect (Hamstra and Feenstra 1989). In order to establish the aspects that are actually considered (and to decide which compact group of objects to select), a pilot study was performed. In this pilot study a collection of application patterns was developed by forming realistic combinations of all aspects and finding existing applications for each pattern. From this list forty applications were selected for the examination of the attitude structure.

The pilot study produced 9 areas of applications in which attitudes were found to differ (Heijs, Midden and Drabbe 1993). The differences were mainly related to the aspects of 'sector' and 'end product/purpose'. Table 1 contains the areas and the applications that were selected for the main study to represent these areas (two applications were included from the controversial area of stock-breeding).

Method and Data Collection

Four surveys were carried out (June 1992, June 1993, June 1994, February 1996). The central part consisted of a series of repeated questions for each area of application covering:

- awareness (heard of the application representing the area);
- attitude (a 7-point scale ranging from highly positive to highly negative);
- 5 cognitive arguments and 2 emotions (positive economic effects; environmental damage; improvement of human health; exceeding of ethical limits; damage to human health;

	Area of Application	Attitude Object	Mnemonic
1.	agriculture: diversification of fruit and vegetables	cell fusion of cabbage varieties to cultivate varieties which can grow in other conditions	cf cabbage
2.	agriculture: product improvement of fruit and vegetables	genetic modification of tomato plants for longer shelf-life and firmness	gm tomato
3.	stock-breeding: new cattle-breeding methods	cloning cattle embryos to breed more calves of a similar quality	cl calf
4.	stock-breeding: new cattle-breeding methods	genetic modification of bacteria using genetic material of a cow for the production of BST	gm BST
5.	veterinary medicine: combating disease	genetic modification of cows using genetic material of humans for resistance to mastitis	gm mastitis
6.	human health-care: production of diagnostics	cell fusion of mouse cells to produce a test of tissue for transplantation purposes	cf transtest
7.	human health-care: production of medicines	genetic modification of rats using genetic material of humans to produce a solvent for blood clots	gm clots
8.	food industry: unfamiliar applications	use of bacteria to produce a raw material to make aspartame	use aspartame
9.	food industry: production of raw materials	genetic modification of yeast using genetic material of a cow to produce chymosine	gm chymosine
10.	chem. industry/ environmental protection: production of raw materials	use of bacteria to produce a raw material for bioplastics	use bioplastics

Table 1. Selection of attitude objects.

enthusiasm; anxiety; 5-point scales ranging from total agreement to total disagreement);
• a positive and negative behavioural intention (to buy or allow the use of the product, or to protest; 4-point scales ranging from definitely to definitely not).

In the third survey perceived control was added (a possible predictor of behavioural intentions). The fourth survey included questions on a general attitude to biotechnology (to investigate whether such an attitude is demonstrable), on basic decision rules or heuristics that might be used to judge unfamiliar applications, on personal experience (as additional predictor variables), and on the cognizance of publications and broadcasts about biotechnology in recent years.

The monitor was subjected to representative random samples of the Dutch population aged 16 and over, ranging from 595 in 1992 to 531 in 1996. Respondents were visited at their homes and the answers were recorded using laptop computers. Special techniques (genetic modification, cloning, cell fusion) were explained in advance.

Knowledge Test

From an original pool of 30 knowledge items (true/false), 15 formed a test with an acceptable level of reliability (a=.62; see Table 2). Seven items are related to knowledge of genetics and biology (marked 'g'), 7 to knowledge of biotechnology (marked 'b') and 1 to legislation (item 12). The test is adequate with regard to the differentiation of difficulty between items, internal coherence and response range. External validity is confirmed by the fact that higher levels of education, inclusion of biology in the curriculum, more interest in biotechnology and higher frequencies of information reception generally go with higher scores on the test.

Results

In this study it is assumed that the data collected on applications will depict attitudes, arguments and emotions linked to the areas of application and thus to other applications in these areas.

1.	Most bacteria are harmful to man (g)
2.	Children resemble their parents because they have the same type of red blood cells (g)
3.	Some bacteria live off petroleum (b)
4.	It is possible to identify Down's Syndrome in a foetus as early as the first month of pregnancy (b)
5.	The hereditary characteristics of living organisms are determined by DNA (g)
6.	The so-called cloning of living organisms produces identical progeny (b)
7.	Viruses can be infected by a bacteria (g)
8.	Genetic modification is the alteration of the hereditary characteristics of living organisms for a particular purpose (b)
9.	Chemical reactions can be speeded up using enzymes (g)
10.	Yeast used in brewing beer consists of living organisms (g)
11.	It is possible to cross a horse with a donkey using normal reproductive methods (b)
12.	It is obligatory to seek government permission before cultivation of a plant whose hereditary characteristics have been changed through genetic modification
13.	It is possible to alter hereditary characteristics of plants to enable them to produce their own pesticides (b)
14.	It is possible to alter the hereditary characteristics of an animal to enable that animal to produce human growth hormones (b)
15.	All the genes of all living organisms on earth are made up of combinations of only four to five different building blocks (g)

Table 2. Selection of knowledge items.

Attitudes

Figure 2 shows the percentages of answers. The range of averages over the years is mentioned below the name of the application. Nine areas are shown since the area of human health-care (diagnostics) was only used in the two surveys with similar results as the area of medicines.

Attitudes are comparable across surveys. The area of environmental protection has the highest scores (with a gradual shift from highly positive to positive). Agriculture and human health-care registered fairly positive attitudes. A neutral response was obtained for the food industry. A marked detail in these areas is the large (and slightly, but non-significantly, increasing) amount of neutral answers, possibly exhibiting a 'wait-and-see' attitude in consideration of the on-coming market introduction. In the case of veterinary medicine opinions were also neutral on average but were more divided. The area of stock-breeding evoked a fairly negative response.

Attitudes to different areas are very diverse. Data of the fourth survey confirm that a general attitude measure is only weakly related to specific applications (ranging from r=.17 to r=.32.) General indicators of attitudes to biotechnology do not appear to reflect attitudes towards a broad range of present biotechnological applications. Therefore the study of more specific attitudes seems a more meaningful approach.

Arguments and Emotions

Correlation and regression analyses show that the relations between arguments and emotions on the one hand and attitudes on the other are almost identical in the four surveys. The formation of attitudes to biotechnology appears to be dominated by emotions, with 'enthusiasm' being the most important. This emotion plays a positive part in most areas but has a negative role in the area of stock-breeding. Anxiety is also relevant for the latter area. The predominance of emotions is in agreement with the theoretical model: attitudes tend to have an emotional basis when there is insufficient knowledge (which is the case here).

Concerning the arguments, the possible exceeding of ethical limits contributes to attitude formation in all areas. This is regarded as

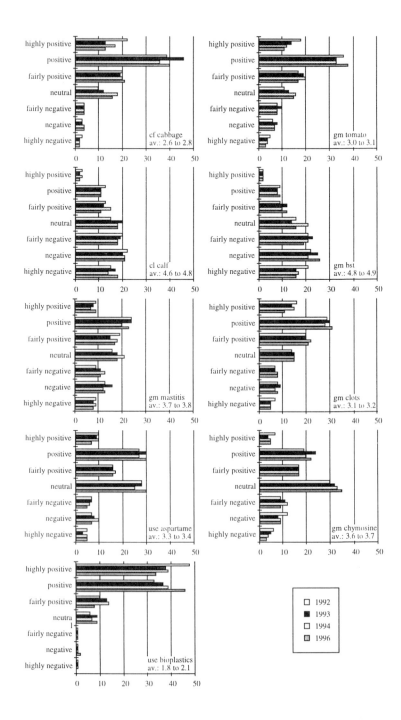

Figure 2. Attitudes relating to areas of application (%).

most likely in stock-breeding and less so in agriculture, environmental protection and the food industry. Other predominant expectations are: positive economic effects from applications in agriculture, the food industry and environmental protection, and positive effects on human health from applications in human health-care.

Knowledge and Awareness

The results of the test indicate a low level of background knowledge (with an average score of about 4 out of 10, corrected for the probability of guessing the right answer). Difficult items concern relatively specialist topics (e.g. the possibility of viruses being infected by bacteria); easier items concern general knowledge on biology, such as the fact that enzymes can speed up chemical reactions. The reliability was reasonable in the pre-test, but showed some variation across surveys (Cronbach's alpha between .65 and .59, but only .50 in 1996). Other test characteristics remained the same (such as the differentiation of item difficulty and the external validity).

Awareness

Respondents' awareness of the various applications was also fairly constant. Applications in stock-breeding and agriculture are relatively well known (by 40% to 65% of the sample). Applications in veterinary medicine, human health-care and the food industry are least known (by 17% to 34%). Environmental protection occupies an intermediate position (38% to 47%).

Interest, Involvement and Personal Experience

During the 4-year period of the study the public remained fairly interested and involved in new developments in biotechnology. This is apparent from the individual significance of the subject, the frequency of listening to and watching of programmes on biotechnology and the amount of general interest. More interest and involvement was in general accompanied by more background knowledge.

The amount of personal experience with applications is expected to influence the formation of attitudes as they become more common in everyday life and new products released. A measure of personal experience was introduced in the fourth survey (ranging from heard or read about it, through seen it or talked to others about it, to used or tasted it and professional experience). Personal experience is still at a very low level and mainly limited to passive information reception. Only aspartame is, to date, recognized to some extent as a product that has been used or tasted.

Sources of Information and Information Needs

Consumer organizations, scientists, the government and environmental groups are regarded as reliable sources. The reliability of the industry, local interest groups, schools and religious organizations is low. To a lesser extent, this applies also to journalism. Journalists are, nevertheless, mentioned as the primary source of information, and current affairs programmes, the news on radio and television, and newspapers as the main information channels. Brochures and discussions with lay people are next in ranking and are more frequently mentioned than might be expected (around 20%). Information from consumer organizations, environmental groups, the government and scientists was both fairly well assimilated and regarded as reliable.

Existing gaps in the information concern: effects on human health and the environment, the controllability of the effects, the legislation, and the limits of scientific capabilities, followed by the quality of new products and the consequences for the Third World. The need for information is considerable: only 10% of respondents regarded themselves as well enough informed.

Since brochures and public information campaigns on television and radio seem to be a fairly important source of information, the fourth survey included a number of analyses on their possible influences. Few significant connections emerged, however. Some of the programmes may have been broadcast too long ago or had a small audience. A brochure on biotechnology and medicines and public debates on the genetic modification of animals and on prognostic genetics drew more attention. Receiving this information is linked to awareness of more applications, greater interest, more background knowledge and a stronger intention to protest against applications in stock-breeding. It is not clear whether this is caused by the information or whether it mainly drew the attention of a public that was already more informed and inclined to protest. The fact that the recipients have higher educational levels points to the latter reason.

Group Differences

In most areas, background knowledge is only weakly related to attitudes. In some areas like the production of chymosine and also the production of bioplastics, weak positive relations are found. The area of stock-breeding shows the opposite: more knowledge was associated with more support of the argument that ethical limits might be violated, less enthusiasm and a less favourable attitude. Apparently, public attitudes may diverge on the basis of interest and information. Persons who consider environmental groups as a main information source are more critical and express more often the expectation of damage to the environment, the exceeding of ethical limits, less enthusiasm and a more negative attitude.

Of the structural variables, gender and age show a consistent relevance in all surveys, especially in the areas of stock-breeding, human health-care and the food industry. Women and elderly people tend to have a more negative argumentation, more feelings of anxiety and a less positive attitude than men and younger respondents. The role of political preference is less constant. In some of the surveys Christian Democrats and Liberals are more enthusiastic in the areas of stockbreeding and veterinary medicine, whereas voters for Labour and Green Left expressed more anxiety, and beliefs that ethical limits might be exceeded and that damage to the environment might occur. Irregular influences of profession and socio-economic status on various arguments are apparent in some cases (higher levels indicate a more positive attitude).

Attitudes and Behavioural Intentions

Behavioural intentions are stable across surveys. Agreement with the positive intention (buying the product or agreeing to its use) is highest in the areas of agriculture, human health-care and environmental protection. It is low in stock-breeding and the food industry. Negative intentions are less frequent: the public is not inclined to overt protest. In addition, a positive (or negative) attitude generally goes with a positive (or less positive) intention whereas the relation between attitudes and negative intentions is less strong. Disapproval is probably more likely to result in a refusal to buy the product or the purchase of an alternative than in political protest. With attitudes as the sole predictors, between 25% and 50% of the variance in the positive intentions can be explained. For negative intentions this is between 3% and 23%. The assumption that the amount of explained variance can be increased by adding perceived control as a predictor (in the third survey) was not substantiated. A possible explanation is that direct confrontations with actual products are still scarce, so that information processing is probably at a level which is too low to incorporate the perception of control in decision making.

Consistent Supporters versus Consistent Adversaries

Consistent supporters were defined as having (highly) positive attitudes to more than 6 applications, and consistent adversaries had a (highly) negative attitude to more than 4 applications. In the third survey men formed a majority among supporters whereas women were predominant among adversaries. In the fourth survey, the moral argument was more important for adversaries in the case of applications in which human DNA is transferred to animals. Adversaries also agreed more strongly with 'conservative' heuristic statements (old techniques suffice, nature should not be tampered with this way), whereas supporters favoured statements concerning the progress of science. Supporters more often mentioned the industry as an important source and adversaries more often pointed to environmental groups. The reliability of scientific, governmental and industrial sources was judged higher among supporters, and the opposite is true for the reliability of environmental and local groups.

Neutral Groups versus (Highly) Positive or (Highly) Negative Groups

In both the third and the fourth survey the largest groups with neutral attitudes are in the food industry. These groups were compared with the groups with (highly) positive and (highly) negative attitudes. The average scores on arguments and emotions in the neutral groups occupy intermediate positions between those in the positive and the negative groups. Background knowledge and interest (1994), personal experience (1996) and awareness (both years) are lower in the neutral group compared to the positive group. Respondents in the neutral and the negative groups differ less with respect to these variables. In addition, the neutral and negative groups have more female respondents (and elderly members in 1994). On the whole, it seems that the neutral groups have more in common with the negative groups.

Conclusions

The study indicates that attitudes towards biotechnological applications vary substantially. The distinction between the nine areas of application appears to be determined mainly by the 'sector' and the 'end product/purpose'. Attitudes to applications in agriculture, environmental protection and human health-care are (fairly) positive. Applications in stock-breeding are judged fairly negatively. Applications in the food industry evoke a neutral attitude and attitudes to applications in veterinary medicine are diverse. This variation creates doubts about the validity of measurements in which the attitude to biotechnology is determined by general questions. Results of the fourth survey confirm that a general attitude is not very meaningful. A compound measurement of public attitudes seems to be indicated.

More positive (or negative) attitudes generally go with more positive (or negative) arguments and emotions. 'Enthusiasm' is the most important determinant of attitudes. Feelings of anxiety have somewhat less influence. From the arguments the moral consideration is most important. The possibility of positive effects on human health comes in second place, followed by the economic argument. The predominant role of emotions is in accordance with the theoretical model: a lack of knowledge prevents the forming of attitudes on the basis of arguments.

Background knowledge and awareness are at a low level and show no improvement over the years. Applications in stock-breeding, agriculture and environmental protection are more familiar and applications in veterinary medicine, human health-care and the food industry are relatively unfamiliar. The higher educated have more knowledge than the lower educated.

Respondents are interested and feel involved. Personal experience is at a very low level. Journalism is the most important source but it is also regarded as relatively less reliable. Scientific sources are seen as reliable but the volume of their information is low. Information from consumer organizations, environmental groups and the government is both heeded and perceived as reliable. The credibility of the industry is low. Radio, television and newspapers are the most important information channels. More information is wanted mainly about effects on health and the environment, the controllability of effects, legislation and the limits of scientific capabilities.

Psychological, social-cultural and structural factors do not relate to strong attitudinal differences, which suggests that the attitudes have not been anchored firmly yet. In the food industry and environmental protection a slightly higher differentiation is visible. The areas of agriculture, stock-breeding, veterinary medicine and human health-care show less differentiation and display an alternating pattern over the years.

Gender and age show some effects. Women and the elderly show a more negative argumentation, more anxiety and a less positive attitude. The roles of political inclination, profession and socio-economic status are less constant.

In most areas, background knowledge is only weakly related to attitudes. In some areas such as the production of chymosine and also the production of bioplastics, weak positive relations are found. The area of stock-breeding shows the opposite: more knowledge was associated with more support of the argument that ethical limits might be violated, less enthusiasm and a less favourable attitude. Apparently, public attitudes may diverge on the basis of the amount of knowledge. This suggests that when people get better informed they will form attitudes which are more certain. The direction of attitudes however is not based on knowledge but on values which are connected to the attitude.

Positive behavioural intentions (buying the product or agreeing to its use) are most frequent in the areas of agriculture, human health-care and environmental protection and less frequent in the areas of stock-breeding and the food industry. Negative intentions are less often mentioned. Relations between attitudes and positive intentions are mostly strong and consistent; they are weaker between attitudes and negative intentions. The public appears to be less inclined to engage in overt protest. Disapproval of products is more likely to result in a refusal to buy the product or in the purchase of an alternative. The theoretical contribution of perceived control in the explanation of intentions was not substantiated. This is probably due to the fact that the level of information processing is too low since confrontations with actual products are still scarce.

The analyses of two potentially relevant population groups for information campaigns lead to the following observations. Consistent adversaries give more weight to the moral argument than consistent supporters if the transfer of human DNA to animals is concerned. Adversaries agree more strongly with 'conservative' heuristics and supporters favour statements concerning the progress of science. Adversaries more often point to environmental groups as a source of information and supporters more often mention the industry. The reliability of scientific, governmental and industrial sources is judged higher among supporters than among adversaries and the opposite is true for the reliability of environmental and local groups.

Groups with neutral attitudes in the food industry show more resemblance to the negative groups as to the positive groups regarding awareness, interest and a predominance of female respondents. This similarity and the fact that women will constitute the majority of potential buyers of these products, should give the neutral groups some priority in information campaigns. The increasing size of the neutral groups is an additional reason for this conclusion.

Reliable information sources with a large range are consumer organizations and environmental groups. Scientific sources and the government are also seen as reliable. Widely used information channels are radio, television and newspapers. Specific television programmes, however, had little influence because they were broadcast too long ago or had a small audience. In contrast, more attention to public debates on biotechnology and brochures lead to awareness of more applications, more knowledge, more interest and a stronger intention to protest. Because of the higher level of education among recipients, a probable explanation is that the publications were seen or read primarily by those who were already more informed and inclined to protest.

Biotechnology can only evolve when public acceptance legitimates the 'licence to operate'. Informing the public may be an important aspect of policy and decision making in the near future. To involve the public it should be reached by use of adequate information channels. The high technological complexity and the low level of knowledge among the public make it difficult for many citizens to form judgements independently. Doubts about moral and health implications may evoke withholding attitudes and a need for control. Trust in regulators and industry may be one of the most crucial factors in the future development of biotechnology.

References

Ajzen, I. 1991. The Theory of Planned Behavior. *Organizational Performance and Human Decision Processes* 50: 179–211.

Breckler, S., and E. Wiggins. 1989. Affect versus Evaluation in the Structure of Attitudes. *Journal of Experimental Social Psychology* 25: 253–271.

Cantley, M. 1987. Democracy and Biotechnology: Popular Attitudes, Information, Trust and Public Interest. *Swiss Biotech* 5 (5): 5–15.

Daamen, D., M. Biegman, C. Midden, J. van der Pligt, and L. van der Lans. 1986. *Individuele oordelen over technologische vernieuwingen.* Report ESC34. Leiden/Petten: Werkgroep Energie- en Milieu-onderzoek/E.C.N.

Durant, J. (ed.). 1992. *Biotechnology in Public: A Review of Recent Research.* London: Science Museum.

Fishbein, M., and I. Ajzen. 1975. *Belief, Attitude, Intention and Behavior: An Introduction to Theory and Research.* Reading, Mass.: Addison-Wesley.

De Flines, J. 1987. Publieke opinie essentieel voor biotechnologie. *De Ingenieur* 6: 6–9.

Hamstra, A., and M. Feenstra. 1989. *Consument en
biotechnologie.* Research Report 85. 's-Gravenhage: SWOKA.

Heijs, W., and C. Midden. 1993. *Biotechnology: Attitudes and
Influencing Factors:* Report for the Ministry of Economic
Affairs. Eindhoven: University of Technology.

Hoban, J., and P. Kendall. 1992. *Consumer Attitudes about the
Use of Biotechnology in Agriculture and Food Production.* North
Carolina: State University.

Knulst, W., and P. van Beek. 1988. *Publiek en techniek:
Opvattingen over technologische vernieuwingen.* Cahier 1988/
57. Rijswijk: Sociaal en Cultureel Planbureau.

De Loor, H., C. Midden, and M. Hisschemöller. 1991.
*Publieksoordelen over nieuwe technologie: De bruikbaarheid van
publieksonderzoek ten behoeve van het technologiebeleid.* Leiden:
R.U., Werkgroep Energie- en Milieu-onderzoek.

Marlier, E. 1992. Eurobarometer 35.1: Opinions of Europeans
on Biotechnology in 1991. In: *Biotechnology in Public: A
Review of Recent Research,* ed. J. Durant. London: Science
Museum.

Meertens, R., and J. von Grumbkow. 1988. *Sociale psychologie.*
Groningen: Wolters.

Petty, R., and J. Cacioppo. 1986. *Communication and Persuasion:
Central and Peripheral Routes to Attitude Change.* New York:
Springer-Verlag.

Tesser, A., and D. Shaffer. 1990. Attitudes and Attitude-Change.
Annual Review of Psychology 41: 479–523.

About the Authors

Lynn Åkesson, Ph.D., Department of European Ethnology, Lund University, has worked on the project 'Transformations of the Body', which resulted in the books *Bodytime* (1996) (edited along with Susanne Lundin) and *Mellan levande och döda: Föreställningar om kropp och ritual* (1997). Since 1997 she has, together with Susanne Lundin, been leading the project 'Genæthnology: Genetics, Genetic Engineering, and Everyday Ethics', financed by The Bank of Sweden Tercentenary Foundation.

Maria Anvret, Ph.D., F.R.C.Path., is Associate Professor and Lecturer in Neurogenetics at the Departments of Molecular Medicine, Genetic Unit, and Clinical Neuroscience, Karolinska Hospital, Stockholm, Sweden. She also functions as a clinical and molecular genetic counsellor.

Carl Reinhold Bråkenhielm is a Professor specializing in studies of faiths and ideologies at the Department of Theology, Uppsala University, Sweden. His publications include *Människan i världen: Om filosofi, teologi och etik i våra världsbilder* (1992), *Livets grundmönster och mångfald: En bok om genetik, etik och livsåskådning* (1995), and *Tro och värderingar i 90-talets Sverige: Om samspelet livsåskådning, moral och hälsa* (1996).

Björn Fjæstad is Adjunct Professor of Science Communication at Mid Sweden University, Östersund. He is also editor and publisher of the Swedish popular science journal *Forskning och Framsteg*. His doctoral degree, at the Stockholm School of Economics, is in Economic Psychology.

Lynn Jane Frewer, Ph.D., is a psychologist working at the Institute of Food Research in Reading, UK. Her research interests include the development of psychological models in the understanding of consumer attitudes to emerging food technologies, developing and testing models of risk perception within the context of food-related risk, effective risk communication strategies, the role of trust and credibility in risk communication, and understanding media reporting of risk, and its subsequent impact on public risk perceptions and behaviours.

Loek Halman is a Research Fellow at WORC, the Work Organization Research Center of Tilburg University, The Netherlands. He is secretary to the Steering Committee of the European Values Study and the board of the Foundation EVS, a member of the World Values Study, and programme director of the 1999 EVS surveys.

Wim Heijs, Ph.D., works as a researcher at the Department of Technology Management, Eindhoven University of Technology. He performs research in the fields of social and environmental psychology, with a special focus on attitudes towards (new) technologies and the mutual influences between physical environments and the users of these environments.

Torben Hviid Nielsen is Professor at the Centre for Technology and Culture, University of Oslo.

Malin Ideland is a doctoral student at the Department of European Ethnology, Lund University, Sweden.

Susanne Lundin, Ph.D., is a Research Fellow at the Department of European Ethnology, Lund University. She has worked on the project 'Transformations of the Body', which resulted in the books *Bodytime* (1996)(edited along with Lynn Åkesson) and *Guldägget: Föräldraskap i biomedicinens tid* (1997). Since 1997 she has, together with Lynn Åkesson, been leading the project 'Genethnology: Genetics, Genetic Engineering, and Everyday Ethics', financed by The Bank of Sweden Tercentenary Foundation.

Alberto Melucci (Ph.D. in Sociology, Ph.D. in Clinical Psychology) is Professor of Cultural Sociology at the University of Milan and Professor of Clinical Psychology at the Post-Graduate School of Clinical Psychology. He has taught extensively at the major universities in Europe, the USA, Canada, Latin America, Asia. He is the author of more than fifteen books. The most recent in English: *The Playing Self: Person and Meaning in the Planetary Society* (1996), *Challenging Codes: Collective Action in the Information Age* (1996), *Nomads of the Present: Social Movements and Individual Needs in Contemporary Society* (1989).

Cees Midden is Professor of Psychology at the Department of Technology Management at Eindhoven University of Technology. His research focus is on man–technology interactions as these become apparent in the development of new products and systems, in societal and market introductions, and in the consumption and use of products. He has published various books and articles on environmental consumer behaviour, on the perception and communication of technological risks and the social diffusion of innovations.

Eleni Papagaroufali, Ph.D., is a lecturer at the Department of Social Anthropology, at the University of the Aegean (Mytilene, Greece). Her present research project is on 'Prospective Donation of Human Organs or Bodies in Greece'. Her previous research includes works on the Greek women's movement, Greek women's cooperatives, and women refugees in Greece.

Katarina Westerlund is working on her Ph.D. in studies in faiths and ideologies at the Department of Theology, Uppsala University, Sweden.